S. HRG. 108–315

# ALIEN SMUGGLING/HUMAN TRAFFICKING: SENDING A MEANINGFUL MESSAGE OF DETERRENCE

# HEARING

BEFORE THE

## SUBCOMMITTEE ON CRIME, CORRECTIONS AND VICTIMS' RIGHTS

OF THE

## COMMITTEE ON THE JUDICIARY
## UNITED STATES SENATE

ONE HUNDRED EIGHTH CONGRESS

FIRST SESSION

---

JULY 25, 2003

---

### Serial No. J–108–30

---

Printed for the use of the Committee on the Judiciary

U.S. GOVERNMENT PRINTING OFFICE

91–788 PDF            WASHINGTON : 2004

For sale by the Superintendent of Documents, U.S. Government Printing Office
Internet: bookstore.gpo.gov  Phone: toll free (866) 512–1800; DC area (202) 512–1800
Fax: (202) 512–2250  Mail: Stop SSOP, Washington, DC 20402–0001

## COMMITTEE ON THE JUDICIARY

ORRIN G. HATCH, Utah, *Chairman*

CHARLES E. GRASSLEY, Iowa
ARLEN SPECTER, Pennsylvania
JON KYL, Arizona
MIKE DeWINE, Ohio
JEFF SESSIONS, Alabama
LINDSEY O. GRAHAM, South Carolina
LARRY E. CRAIG, Idaho
SAXBY CHAMBLISS, Georgia
JOHN CORNYN, Texas

PATRICK J. LEAHY, Vermont
EDWARD M. KENNEDY, Massachusetts
JOSEPH R. BIDEN, JR., Delaware
HERBERT KOHL, Wisconsin
DIANNE FEINSTEIN, California
RUSSELL D. FEINGOLD, Wisconsin
CHARLES E. SCHUMER, New York
RICHARD J. DURBIN, Illinois
JOHN EDWARDS, North Carolina

BRUCE ARTIM, *Chief Counsel and Staff Director*
BRUCE A. COHEN, *Democratic Chief Counsel and Staff Director*

———

SUBCOMMITTEE ON CRIME, CORRECTIONS AND VICTIMS' RIGHTS

LINDSEY O. GRAHAM, South Carolina, *Chairman*

ORRIN G. HATCH, Utah
CHARLES E. GRASSLEY, Iowa
JEFF SESSIONS, Alabama
LARRY E. CRAIG, Idaho
JOHN CORNYN, Texas

JOSEPH R. BIDEN, JR., Delaware
HERBERT KOHL, Wisconsin
DIANNE FEINSTEIN, California
RICHARD J. DURBIN, Illinois
JOHN EDWARDS, North Carolina

JAMES GALYEAN, *Majority Chief Counsel*
NEIL MACBRIDE, *Democratic Chief Counsel*

# CONTENTS

## STATEMENTS OF COMMITTEE MEMBERS


Page

Biden, Hon. Joseph R., Jr., a U.S. Senator from the State of Delaware, prepared statement ............................................................................................................ 34
Cornyn, Hon. John, a U.S. Senator from the State of Texas ............................. 5
Graham, Hon. Lindsey O., a U.S. Senator from the State of South Carolina .... 1
Hatch, Hon. Orrin G., a U.S. Senator from the State of Utah, prepared statement ............................................................................................................................ 72
Leahy, Hon. Patrick J., a U.S. Senator from the State of Vermont, prepared statement ............................................................................................................................ 73


## WITNESSES


Boyle, Jane J., U.S. Attorney, Northern District of Texas, Dallas, Texas .......... 24
Charlton, Paul K., U.S. Attorney, District of Arizona, Phoenix, Arizona ........... 26
Cohn, Sharon B., Senior Counsel, International Justice Mission, Washington, D.C. ...................................................................................................................................... 20
DeMore, Charles H., Interim Assistant Director of Investigations, Bureau of Immigration and Customs Enforcement, Department of Homeland Security ........................................................................................................................................ 6
Harris, Robert L., Deputy Chief, U.S. Border Patrol, Bureau of Customs and Border Protection, Department of Homeland Security ............................ 3
Malcolm, John, Deputy Assistant Attorney General, Criminal Division, Department of Justice ................................................................................................................ 8


## SUBMISSIONS FOR THE RECORD


Boyle, Jane J., U.S. Attorney, Northern District of Texas, Dallas, Texas, prepared statement ............................................................................................................ 36
Charlton, Paul K., U.S. Attorney, District of Arizona, Phoenix, Arizona, prepared statement .................................................................................................................... 41
Cohn, Sharon B., Senior Counsel, International Justice Mission, Washington, D.C., prepared statement ...................................................................................................... 51
DeMore, Charles H., Interim Assistant Director of Investigations, Bureau of Immigration and Customs Enforcement, Department of Homeland Security, prepared statement ...................................................................................................... 59
Harris, Robert L., Deputy Chief, U.S. Border Patrol, Bureau of Customs and Border Protection, Department of Homeland Security, prepared statement ............................................................................................................................................ 65
Malcolm, John, Deputy Assistant Attorney General, Criminal Division, Department of Justice, prepared statement ................................................................. 75

# ALIEN SMUGGLING/HUMAN TRAFFICKING: SENDING A MEANINGFUL MESSAGE OF DETERRENCE

---

## FRIDAY, JULY 25, 2003

United States Senate,
Subcommittee on Crime, Corrections and Victims'
Rights, Committee on the Judiciary,
*Washington, DC.*

The Subcommittee met, pursuant to notice, at 9:55 a.m., in Room SD–226, Dirksen Senate Office Building, Hon. Lindsey O. Graham, Chairman of the Subcommittee, presiding.

Present: Senators Graham and Cornyn.

## OPENING STATEMENT OF HON. LINDSEY O. GRAHAM, A U.S. SENATOR FROM THE STATE OF SOUTH CAROLINA

Chairman GRAHAM. Good morning. I appreciate everyone coming for the hearing. We are going to do something rarely we do in the Senate. We are going to get started ahead of schedule. I do not know if it is going to be a trend that catches on or not but we will try it. The hearing will come to order.

I have a short opening statement. Then I look very much forward to hearing from our witnesses about a topic I think if more Americans were aware of, there would be a lot discussed about what is going on, and I appreciate the people coming today who are on the front lines of fighting this terrible condition that exists in the world.

As we all know, people from all over the world want to come to America to pursue a better life for themselves and their families. Unfortunately, however, some people entrust their lives to some very dangerous people in the effort to gain our shores. I have been told that the business of trafficking human beings is about a $9–1/2 billion business.

Some people brought against their will, kept as human chattel, enslaved in horrible conditions in the midst of our freedom. This is 2003 and people are being taken against their will and forced to work in conditions you would not believe, so the world needs to be more aware that we have not progressed as far as we should have.

I have called today's hearing to examine the issues related to those situations, and after hearing the horrible deaths of aliens smuggled into the country and inhumanely abandoned along a Texas highway last month, I wanted this Committee to examine whether we are doing all we can to combat those horrible crimes. Alien smugglers and traffickers, through unabashed acts of profiteering, endanger lives of countless aliens while compromising the integrity of our immigration laws at the same time. Make no mistake, the human incentives for human smuggling are enormous, $9–1/2 billion.

The commodities involved in this trade, this business, are men, women and children who are smuggled, and they represent substantial profits for those people who decide to do this for a living, and no adjective can adequately describe the people involved in smuggling.

The State Department estimates that more than one million women and children are trafficked around the world each year generally for the purpose of domestic servitude, sweatshop labor or sexual exploitation. At any given time the Department estimates that thousands of people are in the smuggling pipeline with the United States being the primary target. Smugglers deliver some 50,000 aliens here each year.

Alien smuggling is a global problem which requires a systematic and coordinated response. We should do all we can within our criminal laws to combat this terrible problem. I hope this hearing will serve as the beginning of a serious examination of those smuggling and trafficking crimes, and whether our law enforcement authorities have the proper tools to combat them.

Accordingly, we will hear and see the evidence of some of the more tragic stories. I also want to focus attention on those smuggling or trafficking cases where no tragic consequences occur. Unbelievably, some people who traffic human beings are sentenced to time served, months, not years.

Given the risk associated with these crimes every time they are carried out, the punishment should be appropriate to deter future smuggling or trafficking and to sufficiently sanction those who are caught. For instance, the Title 8 provisions provide that a person found guilty of alien smuggling where death results is subject to the full range of punishments including the death penalty.

An issue that I want to explore in this hearing is why when the death results from a Title 18 traffic offense, where the victims are arguably more vulnerable, the defendant is not subject to the full range of punishments, specifically the death penalty. I look forward to hearing from the Department if there is a principle reason why there is a difference in the punishment on those closely-related cases.

In other areas of deterrence we should explore the punishments in those cases where the risk of these serious consequences exists, but because of outstanding law enforcement or efforts, death or serious bodily injury is avoided. In other words, if you are caught involved in the activity where no one is hurt, you have set in motion forces where people could be hurt, and I think we all agree the punishments are too light in those areas.

A debate regarding immigration policy is not what this hearing is about. Senator Chambliss, my good friend, is in charge of the Immigration Subcommittee. This hearing is about the scope of the problem, what we are doing to deter the problem, and what we can do in terms of law enforcement changes to make sure that our criminal laws in this country are sufficient to tackle what I think most Americans feel is a disgusting practice, and what can we do internationally? What can we do to help you, those on the front lines, do a better job in stopping this problem.

I would like to welcome our first panel, and any statements from Senators of the Subcommittee or full Committee, we will introduce into the record at this time. I have a statement of Senator Hatch which will be introduced into the record.

As I said, I would like to welcome our first panel, Assistant Attorney General John Malcolm, who supervises Department of Justice's smuggling enforcement efforts, Interim Assistant Director Chuck DeMore from the Department of Homeland Security, who has the investigative section of the Bureau of Immigration and Customs Enforcement; and finally, Deputy Chief Robert Harris from the Bureau of Customs, Border and Protection in the Department of Homeland Security.

I would appreciate it, gentlemen, if you could limit your statements, if possible, to 5 minutes. I will swear you in. It is just going to be us, it looks like, so we will do this fairly informally, and if you will give me a little bit about the background of your expertise and what your agency does, I very much look forward to your testimony.

If you could rise at this time, please? Raise your right hand. Do you solemnly swear the testimony you are about to give is the truth, the whole truth, and nothing but the truth, so help you God?

Mr. MALCOLM. I do.

Mr. DeMORE. I do.

Mr. HARRIS. I do.

Chairman GRAHAM. Deputy Chief Harris, do you mind starting? Thank you very much.

## STATEMENT OF ROBERT L. HARRIS, DEPUTY CHIEF, U.S. BORDER PATROL, BUREAU OF CUSTOMS AND BORDER PROTECTION, DEPARTMENT OF HOMELAND SECURITY

Mr. HARRIS. Chairman Graham, Ranking Member Biden, and distinguished Subcommittee members, it is my honor to have the opportunity to appear before you today to discuss the mission of the United States Border Patrol, now a part of the Bureau of Customs and Border Protection. My name is Robert L. Harris, and I am the Deputy Chief of the United States Border Patrol.

As you know, on March 1st, 2003, border patrol agents, immigration inspectors, agricultural inspectors, and customs inspectors, merged to form the Bureau of Customs and Border Protection within the Border and Transportation Security Directorate, a part of Homeland Security. Now, for the first time in our Nation's history, agencies of the U.S. Government with significant border responsibilities have been brought together under one roof. With our combined skills and resources, we will be far more effective than we were as separate agencies.

Within the new Bureau the mission of the United States Border Patrol remains unchanged. Our priority mission is the prevention of terrorism, and in carrying out this mission it is our responsibility to patrol our borders between official ports of entry, a mission that is critical to U.S. national security.

Our area of responsibility includes 2,000 miles of the U.S.-Mexico border, 4,000 miles of the U.S.-Canada border, as well as 2,000 miles of coastal waters surrounding the Florida peninsula and Puerto Rico. The current staff of over 10,400 border patrol agents take this mission very seriously. With our increasing capabilities to monitor and patrol—we are increasing our capabilities to monitor and patrol our northern border. In addition to the 245 additional agents for our northern border last year, Commissioner Robert Bonner recently directly the permanent assignment of an additional 374 agents to further strengthen our enforcement presence there. This new deployment, once completed, will provide the American people with a total of 1,000 agents to strengthen security along the U.S.-Canada border. These agents will enforce a strategy with the cornerstones of technology, intelligence, by national and interagency cooperation.

Illegal migration and alien smuggling is a serious problem, and its impact and the associated criminal activity that accompanies it is far reaching. An uncontrolled border presents great concern, spreading border violence and degrading the quality of life in the border communities.

The Border Patrol operates under a comprehensive national strategy designed to gain and maintain control of our Nation's borders. Our operations have had a significant effect on illegal migration along the southwest border, relying upon the proper balance of personnel, equipment, technology and border infrastructure. Our national strategy is based on the concept of prevention through deterrence. Overall, our efforts have been very successful with significant decreases in apprehensions and illegal entries. Apprehensions have declined from a high of over 1.6 million arrests in fiscal year 2000, down to less than 1 million in fiscal year 2002. We are also a leader in the southwest border in narcotics seizures. In fiscal year 2002 our agents seized over one million pounds of marijuana and over 7 tons of cocaine.

In addition to significant arrests of aliens and narcotic seizures, cities like San Diego, El Paso and McAllen have experienced decreased crime rates and an overall improvement in the quality of life. This reduction in crime is due in part to the work of our agents and the effectiveness of our strategy.

Through it all the Border Patrol has encouraged and maintained a positive relationship with local communities and law enforcement agencies, Federal, State and local, operating within the immediate border area. In recent years, unscrupulous alien smugglers have moved migrants into more remote areas with hazardous terrain and extreme conditions. As smuggling tactics and patterns have shifted, our strategy has been flexible enough to meet those challenges. Building on longstanding safety concerns, we have implemented a border safety initiative along the entire southwest border. Striving to create a safer border environment, we pro-actively inform migrants of the hazards before crossing the border illegally, and have established border search, trauma and rescue, or BORSTAR teams, as we call them, to provide quick response to those in life-threatening situations. In fiscal year 2002 our BORSTAR agents recused over 1,700 people in distress. We have developed public service announcements for television, radio and newspaper agencies, both in the United States and Mexico, warning against the dangers of smuggling and illegal entry.

Nationally, the Border Patrol is tasked with the very complex, sensitive and difficult job which historically has presented immense challenges. The challenge is huge, but one which our agents accept willingly, with vigilance, dedication to service and integrity. I know I speak for all of our men and women when I say that we are proud to serve the American people as part of the newly-created Department of Homeland Security.

I would like to thank the Subcommittee for the opportunity to present this testimony today and I would be pleased to respond to any questions that you may have.

[The prepared statement of Mr. Harris appears as a submission for the record.]

Chairman GRAHAM. Thank you, Chief Harris.

At this time I want to welcome Senator Cornyn to the hearing. He has been a tremendous ally of mine in trying to bring the hearing about, and bringing some solutions to the table.

If you would like, Senator Cornyn, you may make an opening statement.

## STATEMENT OF HON. JOHN CORNYN, A U.S. SENATOR FROM THE STATE OF TEXAS

Senator CORNYN. Thank you, Mr. Chairman.

Thanks to all the witnesses who are with us today, and the second panel as well, that we are looking forward to hearing from.

I will just have very brief remarks. I want to say that you all have my profoundest support and empathy, I guess, for the tremendous challenge you have. Unfortunately, I think the challenges are huge and we are not yet—we have not yet met those challenges when it comes to controlling our borders, dealing with exploitation of those who come to this country in order to try to find a better way of life. Of course, in my State of Texas, we have seen recent tragedies dealing with the phenomenon that we are talking about today.

I am grateful to Chairman Graham for convening this Subcommittee hearing so we can talk more about it and what we can do to combat it. I know the facts of the Victoria, Texas case have been widely reported, but I think we can all gain from being reminded of the disturbing facts. In May of this year 17 undocumented aliens were found dead inside of a tractor-trailer. The victims ranged in age from 7-years-old to 91-years-old. These suffered a death while riding in the back of a trailer with possibly over 100 others from Mexico into South Texas. Investigators believe the temperature in the truck exceeded 100 degrees. As of July 17, 2003 a total of 14 defendants have been charged with various smuggling-related crimes arising out of this incident. Despite this case, smugglers continue to use sealed railroad cars and tractor-trailers to move illegal aliens through the South Texas smuggling corridor, and I think we have seen that they are only limited by their imagination in terms of the means and methods by which they bring their cargo, their human cargo across the border. Only days after the discovery in Victoria, Texas, 16 other migrants were discovered in a tractor-trailer only an hour away.

The news of these instances has saddened and angered me, and I think many others. The criminals involved preyed on these families' desire to come to the United States for a better life. I believe these callous and willful actions not only claimed lives and endangered others, but threatened the national security of our country. I understand the Chairman is investigating legislative proposals to increase the penalties for human traffickers, and I applaud him and support those efforts wholeheartedly, and offer my assistance in any way I can possibly provide.

Let me just say in conclusion I appreciate your present here, and you have the support of not only this Committee, but the entire U.S. Congress and the American people, to try to deal with the enormous and sometimes it seems overwhelming challenges we have when it comes to regaining security of our borders and dealing with the profoundly sad and tragic exploitation of those who cross our borders seeking a better life.

Thank you, Mr. Chairman.

Chairman GRAHAM. Thanks, Senator Cornyn.

Mr. DeMore?

## STATEMENT OF CHARLES H. DEMORE, INTERIM ASSISTANT DIRECTOR OF INVESTIGATIONS, BUREAU OF IMMIGRATION AND CUSTOMS ENFORCEMENT, DEPARTMENT OF HOMELAND SECURITY

Mr. DEMORE. Thank you, Mr. Chairman, and members of the Committee. I am Chuck DeMore, Interim Assistant Director of Investigations for the Bureau of Immigration and Customs Enforcement, and I thank you for the opportunity to address you regarding our efforts to combat smuggling of aliens into the United States.

The creation of the Department of Homeland Security and specifically BICE, combined the legal authorities and the investigative tools necessary to effectively combat organized human smuggling and trafficking.

I would like to begin by providing an important clarification and necessary distinction between the terms "alien smuggling" and "human trafficking." Human trafficking involves force, fraud or coercion, and occurs for the purpose of either forced labor or commercial sexual exploitation, generally over an extended period of time. Alien smuggling, on the other hand, is an enterprise that produces short-term profits resulting from one-time fees paid by the smuggled aliens themselves or their respondents. Human smuggling has become a lucrative international criminal enterprise and continues to grow in the United States. The commodities involved in this multibillion dollar illicit are men, women and children.

Traffickers and smugglers transport undocumented migrants into the United States for work in licit and illicit industries. The trafficker's foremost goal, like the smuggler, is to maximize profits. To illustrate the callous disregard smugglers have for human life, I would like to provide you with the details of some recent tragic smuggling involving deaths.

In October 2002 in Iowa, 11 undocumented aliens were found dead in a covered grain car near Dennison, trapped in the grain car for 4 months. The crime is the subject of an ongoing investigation.

In May 2003, as was alluded to, in Victoria, Texas, 17 undocumented aliens were found dead inside a tractor-trailer. Four hours into their 300-mile trip to Houston, oxygen ran out in the dark, sealed, hot and airless trailer. The trapped aliens beat their way through the trailer tail lights in a desperate attempt to signal for help. Within 72 hours of the grisly discovery, special agents and intelligence analysts from BICE, in collaboration with our colleagues from the Bureau of Customs and Border Protection, the Texas Department of Public Safety, the Victoria County District Attorney's Office, the United States Secret Service, and the Victoria County Sheriff's Office, identified and arrested four defendants in Ohio and Texas. As of July 17, 2003, a total of 14 defendants have been charged with various smuggling-related crimes arising from this tragic incident.

Still, as you suggested, Senator, smugglers remain undaunted by this tragedy. They continue to use sealed railroad cars and tractor-trailers to move aliens throughout the South Texas smuggling corridor. In fact, only days, as you suggested, after the discovery in Victoria, 16 other migrants were discovered in a tractor-trailer an hour away. It remains unknown what their fate might have been had they not been discovered by law enforcement.

In January 2000 in the State of Washington, 3 undocumented aliens were found dead in a cargo container in Seattle. The 3 were part of a group of 18 smuggled Chinese aliens that had been sealed in the container for a period of 2 weeks crossing the sea. The survivors, who were in dire medical condition, remained in the container with the deceased until they were discovered.

In March 2000 in California, 6 undocumented migrants were found in the San Diego East County mountains, 4 of whom died due to hypothermia. The smugglers abandoned the group in the snowy mountains even as the aliens pleaded not to be abandoned.

In December 2001 in Florida, a capsized vessel was found in the Florida straits, alleged to have been carrying 41 Cuban nationals including women and children. All are believed to have perished at sea.

Finally, last year in Arizona, 133 deaths were attributed to alien smuggling in the unforgiving Arizona deserts. Tragically, many of these deaths were due to aliens being abandoned in the desert heat. Unfortunately, not all of the deaths were accidental. The BICE Special Agent in Charge in Phoenix is currently investigating several alien-smuggling organizations believe responsible for 13 homicides. Several of the deceased were undocumented aliens who were unable to pay for their smuggling fee, and instead were forced to pay with their own lives.

I'm pleased to note that just yesterday, some very significant arrests were made in Phoenix with respect to these ongoing investigations. Local law enforcement agencies in Arizona attribute most of the increase in immigrant-related violent crime, hostage taking and home invasions, to an active alien smuggling trade there.

It is clear as evidenced by these examples that alien smuggling is not confined to any geographic region, is a problem of national scope which requires a coordinated national response. BICE is currently developing a foreign and domestic anti-smuggling strategy, which has as a cornerstone the implementation of critical incident response teams. The purpose of these investigative teams is simple and effective, to begin the investigation, like in Victoria, of a critical incident as quickly as possible, bringing to bear this Bureau's broad spectrum of statutory authority supported by a robust infrastructure to include language and cultural specific interview skills, land, air and marine smuggling assets and intelligence crime scene management, victim, witness, forensic and financial tracking specialists.

Increased efforts are also placed on addressing the smuggling of juveniles into the United States which has surged in recent years. This increase is driven by the demand created by U.S. citizens and others wanting to illegally adopt children from abroad, immigrants attempting to reunite their families, and various forms of child exploitation. Trafficked children are often lured by promises of education, a new skill or a good job. Other children are kidnapped outright and then bought and sold as commodities.

In the aftermath of the September 11th attacks we also more fully appreciate our vulnerabilities with respect to terrorists using established smuggling networks to threaten domestic public safety, as well as American interests abroad. BICE views the destruction of transnational smuggling and trafficking organizations as an extraordinarily high priority.

The most effective means of addressing this vulnerability is by attacking the problem in source and transit countries, thereby preventing the ultimate entry of our enemies into the United States. To that end we presently serve as co-chair with the Department of Justice and with Central Intelligence Agency to an interagency working group targeting smuggling organizations that present national security concerns for the United States.

Members of this Subcommittee have previously raised the issue of a need for enhancing penalties for smuggling offenses. While we believe that penalties set forth in Section 274 of the INA to be adequate, in practice, the sentences imposed in other than the very high-profile cases traditionally have been quite short.

We look forward to working with this Committee in our efforts to save lives and secure national interests. I thank you for inviting me to testify, and I will be glad to answer any questions.

Time permitting, I do have some compelling photographs which depict the callous disregard alien smugglers have for their human cargo, which at your convenience, we would be happy to show. Thank you, sir.

[The prepared statement of Mr. DeMore appears as a submission for the record.]

Chairman GRAHAM. Thank you.

Mr. Malcolm?

## STATEMENT OF JOHN MALCOLM, DEPUTY ASSISTANT ATTORNEY GENERAL, CRIMINAL DIVISION, DEPARTMENT OF JUSTICE

Mr. MALCOLM. Mr. Chairman, members of the Subcommittee, thank you for the opportunity to appear on behalf of the Department of Justice to discuss the problems of international alien smuggling and human trafficking, which we sometimes to as trafficking in persons.

These two serious crimes, distinct in nature, but related in their effects, are of great importance to the Department because they present both national security and human rights concerns. In my role as a Deputy Assistant Attorney General in the Criminal Division, I oversee the Domestic Security Section and the Child Exploitation and Obscenity Section, which are the offices within the Division that focus on these offenses.

The Department's Civil Rights Division also has criminal prosecutors who target human trafficking, and we often coordinate our efforts with them in combatting these pernicious offenses.

Ultimately, alien smuggling and human trafficking subvert our Nation's sovereignty. Alien smuggling puts the decision about who enters our country into the hands of criminals who may not know and probably don't care if their actions help terrorists or other criminals to enter our country. Both types of crimes strain limited resources and penalize persons who wish to enter our country legally. These crimes also enable international criminal organizations to flourish throughout the world and breed corruption often of border officials in other countries, thereby undermining respect for the rule of law and harming basic democratic institutions.

Some have argued that alien smuggling is a so-called victimless crime. I would like to put that pernicious myth to rest. Smuggled migrants are often subjected to violence and inhumane and dangerous conditions. Some are trafficked into sexual exploitation or forced labor. Others die every year from drowning, abandonment, accidents of brutality by smugglers. While it is not uncommon to find one or two bodies in the mountains or in the desert, it's only when a large number of migrants die that the national attention focuses on the danger that these desperate people face. The recent deaths of nearly a score of migrants in Victoria, Texas who suffocated to death in a tractor-trailer, abandoned by their smugglers, is only the most recent example.

Even if migrants arrive at their intended destinations alive and unscathed, however, smugglers have been known to extort payments and exorbitant fees by forcing migrants into virtual slavery, including selling them into sexual exploitation or by holding family members, back in their home country, for ransom. The Department takes these cases seriously.

Primarily through United States Attorneys' Offices around the country, we prosecute a large number of alien smuggling cases every year. In 2001 we obtained approximately 1,900 convictions for alien smuggling offenses, 17 percent of the total number of immigration related convictions for that year.

We owe this Subcommittee, the Judiciary Committee and the Congress as a whole a debt of gratitude in enacting legislation that has helped us fight immigrant smuggling and human trafficking. Most notable were the Trafficking of Victims Protection Act of 2000, as well as legislation in recent years that focused on alien smuggling and document fraud crimes and penalties. Relying on the TVPA and a variety of other statutes, the Department has moved forcefully to punish traffickers as well as to assist their victims. As of March 2003, there are 128 open trafficking investigations, nearly twice as many as there were in January of 2001.

We use various methods to investigate and prosecute these crimes. Sometimes agents discover smuggling operations or human trafficking operations while in progress, as occurred in Victoria, Texas. More typically, especially in smuggling cases, law enforcement officers will encounter a small group of persons along the border with a coyote, who is the person who has been designated as a guide to guide the migrants over the border. Sometimes agents will receive informant tips or intelligence from foreign countries. Particularly in cases of large-scale criminal operations, those investigation may be long term, resource intensive and involve the use of confidential informants and undercover agents.

As with the war on terrorism, interagency and international cooperation is essential to our efforts. The Criminal and Civil Rights Division work with other agencies such as the FBI, the Department of Homeland Security, the Labor Department and the State Department, as well as with foreign law enforcement counterparts.

We also have worked to strengthen the laws of other countries and to make smuggling and trafficking extraditable offenses. We have assisted other countries in their efforts to remove corrupt officials who aid alien smuggling organizations, and have helped our foreign counterparts to initiate their own prosecutions.

Lastly, I would like to describe a couple of recent cases that exemplify our alien smuggling and human trafficking prosecutions. Last year here in Washington we tried and convicted an Iranian national, Mohammed Hussein Assadi, for alien smuggling. Assadi ran a large-scale organization that smuggled aliens, generally from Middle Eastern and South Asian countries, into the United States on commercial airlines. Assadi's ring provided aliens with stolen photo-substituted passports from European countries that qualify for visa waiver privileges under U.S. laws. As a result of interagency and international cooperation from other countries, Assadi was apprehended and then expelled from the foreign country, and he was sentenced to 30 months in prison.

A recent human trafficking case, United States v. Kil Soo Lee, involved sweatshops in American Samoa. FBI agents worked with the Labor Department and INS investigators to uncover a trafficking scheme where 200 Vietnamese and Chinese nationals, mostly young women, were smuggled into American Samoa from Vietnam to work as sewing machine operators in a garment factory. The traffickers held these women for up to 2 years, using extreme food deprivation, beatings and physical restraint to force them to work. In February a jury convicted Kil Soo Lee, the owner of the factory and leader of the organization on nearly all of the counts. Two other defendants pled guilty. Mr. Lee is going to be sentenced this December and faces a substantial prison term.

The sentences in smuggling and trafficking cases have varied significantly. Some cases have had significant sentences and a number have not. It's my understanding that the Sentencing Commission has put on its agenda for the upcoming year a review of immigration related guidelines, and the Department looks forward to working with this Subcommittee and with the Sentencing Commission on these issues. In particular, as noted in my written testimony, and as will be, I expect, highlighted by some of my colleagues, such as Paul Charlton in the second panel we believe that sentences, particularly in alien smuggling cases that have not resulted in death or violent injury do not always appropriately reflect the seriousness of the crime, and we look forward again to working with the Subcommittee to improve our laws.

We believe that this could assist the Government's efforts to discourage illegal immigration while prosecuting to the full extent of the law those smugglers and traffickers who deal in human misery, exploitation and desperation.

I would ask that the full text of my written comments be included in the record, and with that, I would be happy to answer the Subcommittee's questions.

[The prepared statement of Mr. Malcolm appears as a submission for the record.]

Chairman GRAHAM. It will be ordered as such.

Could we see the photos right quick? Then we will go to Senator Cornyn.

Mr. DeMORE. I believe one of my two colleagues at the table might have personal involvement in these cases as well, so, please, if you do, join in.

This just shows the disregard, as I suggested before, that the alien smugglers show for the aliens. This individual is secreted in a dashboard of a vehicle.

Chairman GRAHAM. That is a person in a dashboard?

Mr. DeMORE. This is a woman, whose face you can see if you were to open the glove box, but her body literally is secreted behind the dashboard.

This gentleman here was literally—he had a chair built around him, and you can imagine the heat in that configuration. And as you know, the line sometimes getting into the country is significantly significant, so.

This just shows over-the-road smuggling. The number of people that are put into some of these vehicles are beyond comprehension. I've seen 25, 30 people come out of a vehicle that were really—

Chairman GRAHAM. What will they charge someone?

Mr. DeMORE. Now, when I was doing alien smuggling, the fee was $300 from roughly the Tiajuana area to Los Angeles. I am told now because of the successes of the Border Patrol on the border, it is closer to $1,000 to $1,500 for the—

Senator CORNYN. You mean when you were combatting it?

Mr. DeMORE. Yes, sir.

Senator CORNYN. I think you said when you were doing alien smuggling.

[Laughter.]

Senator CORNYN. We understood what you meant, but I wanted the record to be clear.

Mr. DeMORE. Again, you see the number of people that are put into these vehicles put them in great harm.

Senator CORNYN. How old are those children?

Mr. DeMORE. You will find people put in trunks and backs of vehicles that children, elderly people, there is no real regard for their physical comfort or safety. These two vehicles here were subject to rollovers. Very often the smugglers will ask one of the smuggled aliens to drive for a reduced fee, and people who have very limited skills, and particularly on the highways in the United States take behind the wheel, and often they result in rollovers and fatalities.

Senator CORNYN. Go ahead.

Mr. DeMORE. Maritime smuggling is also a significant threat. Hundreds and hundreds of people may be put on a vessel, three or four weeks at sea. We have found aliens have been thrown off of the vessels. If, for whatever reason, they seem to pose any kind of threat to the smugglers, they have been known to be thrown into shark-infested waters. We have actually recovered bodies that have been attacked by sharks after the smugglers had thrown them overboard.

Use of containers, as the testimony suggested, we have seen that coming from China—people put for weeks on end into containers and obviously some perish and the rest are literally trapped in the container with the deceased until such time as they can be—

Chairman GRAHAM. And you are talking about weeks, not days.

Mr. DeMORE. Weeks, yes, sir.

This is the Victoria case. Obviously, ultimately, 19 deceased aliens. In fact, it is so horrific that I will tell you that interviews of the aliens suggested that a group of adult males wished to dismember the child that you referred to Senator so as to be able to force his body parts through the truck—so as to alert people on the outside that they were in such peril. That is the degree of, that is just horrific. Obviously, the bodies. These are the 14 individuals that have been taken into custody, charged with various crimes relating to this conspiracy.

Now, these people are, are they the masterminds or are they the worker bees or do we know?

Mr. DeMORE. It is a combination of both.

Chairman GRAHAM. A combination of both.

Mr. DeMORE. We have some significant players here, a woman that would, immediately after the arrest, went to Honduras. We do not have extradition with Honduras. We ultimately were able to lure her into Guatemala and ultimately take her into custody.

Chairman GRAHAM. What do we need to do to get extradition with Honduras?

Mr. DeMORE. I sure the Department of State could probably best answer that, but that would be helpful. We have known of other cases, and I think one of the U.S. attorneys will refer to that in the second panel.

This shows some of the smuggling routes. Smugglers will use multiple transit countries, numerous conveyances, types of conveyances, travel documents from a whole host of countries, some good, some bad.

This is a proprietary. This is actually a house that we set up in the context of an alien smuggling organization so we could identify the alien smugglers and subsequently present for prosecution. It was all set up for video and audio. We could actually watch the smugglers bring the smuggled aliens into the house and identify who was involved, and what kind of capacity, in this operation.

This is the monitoring of the load house from an adjacent residence, where we had the tech equipment set up.

This is a really interesting vehicle. This was interdicted last week by the Coast Guard. It is an old Chevy truck that they affixed pontoons to, configured the drive train to turn a propeller, and I am told that it was about 50 miles at sea when it was interdicted by the Coast Guard. And my first question was that is interesting that they were wearing life rafts, but those were obviously dropped by the Coast Guard.

So this is the kind of ingenuity the smugglers will use and the dangers, the present dangers, they put the aliens, subject them to. I think that is it.

Chairman GRAHAM. Thank you very much. It is very informative. Senator Cornyn?

Senator CORNYN. Just so we understand the scope of the problem, a couple of you I guess have differentiated between human trafficking and alien smuggling, and I assume those are both done for a profit motive; is that correct, Mr. Malcolm, in your experience?

Mr. MALCOLM. Yes, Senator. The difference between the two, very simply, is that people who are trafficked are brought in by force, fraud or coercion, and once they reach their destination, they are enslaved in either domestic labor, farm labor or sex labor situations. People who are smuggled have paid a fee for the purpose of coming into this country, and once they pay those fees, often being exorbitant, they are set loose into the community or they join loved ones to make it on their own, but they are both highly profitable activities.

Senator CORNYN. But I assume that if someone is willing to, for money, traffic in human beings or to assist in the smuggling of human beings, there is no reason why that same person, interested in purely a profit motive, would not cooperate with terrorists in infiltrating our country. Would you agree with that Mr. DeMore?

Mr. DEMORE. Yes, sir, I would. In fact, we know of Middle Eastern nationals who are resident in third-party countries, Ecuador, Uruguay, other countries, that actually facilitate the movement of aliens from the Middle East through transit countries destined to the United States.

Senator CORNYN. Well, for example, in my State of Texas, we share a 1,200-mile common border with Mexico, and some have concerns, and I share those concerns about the consequences of human smuggling, when it is people who are coming across the border merely to work here, and what happens to them, but I also have a concern about others from other countries who may come to Mexico and across our borders because they know that that is a porous border, and getting less porous all of the time, I hope, but still porous nonetheless.

Mr. Harris, in your experience or your agency's experience, have you seen that phenomenon?

Mr. HARRIS. Could you clarify the question, please.

Senator CORNYN. Sure. People coming from other countries to a place like Mexico to cross our border on the U.S.-Mexican border, rather than entering through some other route because they know that that is a vulnerable border to trans-border crossings.

Mr. HARRIS. Certainly. As I stated in my opening statement, Senator, we arrest on average over a million people a year, by and large the bulk of those, over 90 percent of those are from the country of Mexico, but the people that we encounter from other countries, you can just name pretty much any country in the world, we arrest people from those areas.

We have, this year, arrested about 250 aliens that we refer to from special interest countries where there is a high degree of interest by the U.S. Government. The bulk of those arrests occurred on the U.S. Canada border, but these were not, you know, the majority of them were not people actually crossing the border. They were encountered by other law enforcement agencies during jail check and what have you.

Senator CORNYN. So those apprehensions could include countries that are state sponsors of terrorism on the State Department's Watch List, for example?

Mr. HARRIS. Certainly. But, by and large, along the U.S.-Mexico border, we have not encountered any significant increase or any significant numbers of those people crossing the U.S. and Mexico border. Certainly, we do recognize it as vulnerability, but in terms of apprehensions and what have you, we have seen nothing significant, no.

Senator CORNYN. You catch about a million a year?

Mr. HARRIS. On average about a million a year. In fiscal year 2000, 1.6 million; 2001, 1.2 million; last year was 970,000/980,000, around in there. We expect a similar number this year.

Senator CORNYN. Do you know how many make it across that, due to lack of resources or whatever the reason may be, technology, that you do not catch?

Mr. HARRIS. Well, there is all kinds—

Senator CORNYN. Or what percentage you do catch?

Mr. HARRIS. I do not have a good answer for that question. I will tell you that right now. I have heard and seen some of the estimates. I will tell you that in areas where we do have adequate resources, where we have implemented our strategy, where we have sufficient personnel equipment and technology, we are much better at being able to come up with that number. Especially where we have the camera systems and sensors and what you have, we can tell pretty much what comes across the border and then account for what we catch and what we don't. But in areas where we don't have those kind of resources, we don't have a good way to get that number.

Senator CORNYN. I have seen figures, it seems to me, and I am not trying to be precise here, but somewhere on the order of hundreds of thousands that still come across our borders that are not apprehended. Would you have any reason to disagree with a number in the 200,000 range?

Mr. HARRIS. I would have to look at those figures and how someone would come up with those types of estimates.

The figure that I see most recently referred to is the Census data about how many illegal aliens we have in the country, although I would say that those do not necessarily represent people who cross the border between the border between the ports of entry and escape. A lot of those country came into the country legally, through the ports, with proper documents and did not leave the country. I understand that the Department has been working diligently to establish a means of tracking those types of people to make sure that they do leave the country.

Senator CORNYN. The figure I have seen is about 40 percent of our illegal immigration is a result of people who have overstayed their visas, which would leave, I guess, 60 percent who have gotten here without any legal authority whatsoever. But I guess it is really a hard question to say you know who you catch, but I guess you do not know who you do not catch.

Mr. HARRIS. Certainly. Very difficult to measure. We know, obviously, we're not catching everybody. We think we're getting much better at it, but obviously we do have some work left to do in a lot of areas, especially along the Southwest border.

Senator CORNYN. The figures I have seen are we have between 8 to 10 million undocumented immigrants in this country. I was shocked to learn, with Senator Chambliss in his Subcommittee last week, that we have about 300,000 people currently in the United States under final orders of deportation. They have exhausted all legal remedies, and we simply do not know where they are. So the problem is huge in magnitude.

Two other quick subjects I wanted to ask about, and perhaps, Mr. Malcolm, this one would relate to you. Chairman Graham has called this to look at what sort of penalties the Congress might enact or consider to deter this sort of reprehensible activity. But we have a problem, I guess, particularly with Mexico and other countries that simply will not extradite their citizens if they are subject to a death penalty. I believe that also extends to those who have potentially a life sentence. Is that correct?

Mr. MALCOLM. I believe so.

Senator CORNYN. So that is certainly, I guess, Mr. Chairman, something we are going to have to consider and certainly something that the State Department in our discussions with our neighbors in Mexico, if we are going to ratchet up the penalties, whether we are going to get the kind of cooperation that I believe we need with neighboring countries and other countries in terms of extraditing those criminals to this country for proper punishment.

Finally, let me just ask, Mr. DeMore, a couple of weeks ago I filed a bill called the Border Security and Immigration Reform Act of 2003, which I hope will restart the discussion about a guest worker program in this country. I simply want to ask you: If Congress is able to come up with a legal means for people to come from other countries to this country to work and then to return to their home country with the money that they have earned, will that relieve some of the pressure that we feel now because people feel like they have no means to get here legally, so they simply turn themselves over to human smugglers and others who have no concern for them at all?

Mr. DeMORE. Sir, if the Congress were to be so inclined, we would work with the Congress to make sure that whatever methodology was imposed would be—we would safeguard the national security interests in the context of working with you.

Senator CORNYN. And I do not mean to put you on the spot to ask you to endorse some legislation, but it just makes sense to me, Mr. Chairman, if there was some legal means for people who wanted to come here and work and then return to their country, then there would be less people who would turn their lives over to these human smugglers and suffer perhaps death and horrible injury in the process, because they really feel like they have no other way to get here.

Thank you very much.

Chairman GRAHAM. Thank you, Senator, and I would like to associate myself with your comments about trying to create some order out of chaos. The market forces are what we are talking about. There are many industries in this country that rely on the illegal alien workforce, the immigrant workforce. There are many jobs in America that are being filled by this demographic group, and if we could find a way to make it so that you could come and it would be a win-win, help the American economy, help the individual involved lawfully enter the country, lawfully leave, save some lives, I think that is a must before we can ever get our hands around this problem.

But what I want to talk about is in 2000 we passed the Victims of Trafficking and Violence Protection Act, which I think is a great thing to have done. The reason I wanted to have the hearing is to sort of inventory where we are at, what can we do to supplement your efforts to enforce this Act, what deficiencies have we seen in the Act, and sort of give you a chance to make a shopping list. If you could, what would you change about it?

The first question I have, Mr. Malcolm, is: Human trafficking, I think I understand after your testimony, is different from alien smuggling. Alien smuggling, you take money and your job is to get them here. And once that job is complete, they sever the connection. The way you make your money in human trafficking is you take somebody against their will, and you use their body or their services to make money. You make slaves out of them, you sexually exploit them or other things to get money out of them.

Does the death penalty apply in a situation where somebody is involved in human trafficking and a death results?

Mr. MALCOLM. Actually, Senator, in terms of traffickers, you can make money both ways. You can promise—get people here by fraud, giving them false promises and extort a fee from them, and then get them here and enslave them.

Chairman GRAHAM. Right.

Mr. MALCOLM. The answer to your question is no, there is no death penalty provision for human trafficking.

Chairman GRAHAM. Would you suggest that we embark upon making that a death penalty offense if someone dies as a result of that activity?

Mr. MALCOLM. I wouldn't want, Senator, to commit prematurely on behalf of the administration. However, as you said in your opening statement, you weren't sure why—whether there was a principled reason for a difference between the two, and I can't see one either.

Indeed, both alien-smuggled victims and traffic victims are subjected to horrendous risks while being transported here. Once they get to the destination countries, alien-smuggled victims are frequently set free; whereas, for traffic victims, the danger and horrors for them are just beginning.

Chairman GRAHAM. I am just speaking for myself. It seems between two classes of cases, the people that would be the best candidates for the death penalty are someone who seizes by force or coercion or trickery and enslaves someone and makes their money from abusing their body or abusing their services. And if you could get back and check with your superiors about that, I would like to have this Committee looking into changing—I guess it's Title 18?

Mr. MALCOLM. Yes, that's correct.

Chairman GRAHAM. And make sure that the death penalty applies in both situations.

Mr. MALCOLM. We'd be happy to work with you on that, Senator.

Chairman GRAHAM. Thank you.

Mr. MALCOLM. The only thing I would say is that with respect to smuggling organizations as opposed to traffickers, those are the routes that are used by terrorists so they service somewhat different audiences, but they're equally pernicious crimes and we're happy to work with you.

Chairman GRAHAM. I appreciate that very, very much.

Mr. DeMore, Senator Cornyn sort of asked a question about legal changes in terms of how people come, to have a route to get people here where they can actually help the economy and return to their home country and it would be a win-win. What is your opinion of an immigration change where people would have a route to come to America for the purpose of working temporarily, then going back to the country of origin?

Mr. DeMORE. Sir, I think that's a dialogue that should certainly take place, and we would be most happy, as I said before, to engage you in that discussion and do so in a way that would be consistent with ensuring the integrity of the American immigration process.

Chairman GRAHAM. Now, this is a global problem, as you have demonstrated. Is there any particular country that is more friendly to this type activity than others? Is there a particular regime out there where, if you are a human trafficker or a human smuggler, this is a good place to work from?

Mr. DeMORE. I wouldn't look at a particular country from the context of the government providing any kind of safe haven, but there are areas where foreign government officials are sometimes exploited by the alien smugglers, and anywhere where there is a deteriorating economy, government officials are more subject to being corrupted and lends itself to the basic kinds of places where you would expect to see alien smugglers operate relatively freely, certain countries in Asia, South America, Central America, the Caribbean.

Chairman GRAHAM. Chief Harris, if we move people to Canada, the main reason we are doing that is to try to prevent people who are of the terrorist mentality from having easier access?

Mr. HARRIS. As I mentioned in our opening statement, our mission, the mission of Border Patrol, has remained virtually unchanged. Our responsibility is to patrol the border between the ports of entry. But our priority mission within that and part of Customs and Border Protection is the prevention of terrorism and entry of terrorists and weapons of mass destruction.

Chairman GRAHAM. My question is: Generally speaking, Canadians are not flooding the borders. Canadians are not coming across in great droves. Why are we moving people to the Canadian border?

Mr. HARRIS. The U.S.-Mexico border is 2,000 miles. The U.S.-Canada border is 4,000 miles, so it is twice the length, a lot of vast, open terrain. Prior to the attacks of September 11th, we had about

368 Border Patrol agents who were responsible for patrolling that 4,000 miles, certainly inadequate, but our strategy was to put the bulk of our resources where we saw the majority of the illegal immigration problem, which was along the Southwest border. But certainly the attacks of September 11th changed our Nation and our agency. You know, it's never going to be the same again.

So we recognized fully that the Northern border, even though we didn't have the volume, it represented a vulnerability and that we needed to do something to close that gap. Congress passed the PATRIOT Act stating that we needed to triple the amount of resources on the Northern border. We certainly intended to do that. Last year, we put 245 additional agents up there. Commissioner Bonner recently directed us to deploy additional agents. We think those thousand agents, together with additional technology that we are putting up there, the camera systems and the sensors, is going to provide us with a degree of detection and response capability along that 4,000-mile border that we never did have before.

Chairman GRAHAM. Was the border a place where people involved in human trafficking would come down from Canada? Has that ever been a serious problem in the past?

Mr. HARRIS. I have seen and read some reports about, for example—I'll give you an example—saying that the cost of an airline ticket from Mexico to Canada was less expensive than paying a smuggler $1,000 or $2,000. So the possibility existed that, you know, people were flying up to the Canadian—into Canada and then crossing the U.S.-Mexico—or U.S.-Canada border. But I have not seen that. The statistics, the intelligence that we have is—

Chairman GRAHAM. There is no data to suggest that—

Mr. HARRIS. Does not prove that out.

Chairman GRAHAM. Right.

Mr. HARRIS. On average, we make about 13,000 arrests a year along the Northern border, but about 70 percent of those are aliens who crossed the U.S.-Mexico border and then migrated up North and they were apprehended by our agents on patrol there. So not a lot of people crossing, but a very long border that needs to be patrolled. I mean, it does represent a vulnerability. We're trying to shore that up a bit.

Chairman GRAHAM. Well, the last question I have I will address to each of you. We are a couple of years into the Victims of Trafficking and Violence Protection Act, and if you could, just as briefly as possible but not cutting yourself short, tell us how you believe the Act is working and what would you like to see changed. And I know you may not give us a full answer now, but what I would suggest is go back to your agency, inventory among yourselves, sit down and think about it a while, and kind of give us a list. What would you like to see us do in terms of legal changes or structural changes to make this Act more effective? And that is an open invitation.

Mr. Malcolm, could you tell us generally how the Act is working and what you would like to see changed, if anything?

Mr. MALCOLM. In terms of the changes, Senator, we'll get back to you. In terms of how it's working, the answer is, in short, very well. The Act doubled sentences that were potentially available, made immigration benefits, medical benefits, available to victims of severe forms of trafficking. We have more open cases. The hotline is working extremely well. The Act provided for an annual report by the State Department, the tracking persons report, which ranks countries. Countries that are so-called tier three countries have to improve their anti-trafficking effort or face potential assistance sanctions, and that's provided a great incentive for them to cooperate with our efforts. The Act is working very well.

Chairman GRAHAM. Mr. DeMore?

Mr. DEMORE. I agree with my colleague. We've received over 450 applications. We've approved 172, 238 pending. I'm not aware of any immediate needs for structural change to make it better, but I'll certainly dialogue with my colleagues that are involved in this daily and see if we can come up with some suggestions.

Chairman GRAHAM. And if you could give us some information about sentencing behavior, that is very important to me because I want to talk with the Sentencing Guideline folks and make sure that we are addressing what I think appears to be a deficiency, that the average sentences for many of these cases without injury is very, very low, and you are not taking the incentive out of doing this. If you could do that, I would appreciate it.

Chief Harris?

Mr. HARRIS. A couple of comments, Mr. Chairman. First, I appreciate the opportunity to go back and take a look at it and come back to the Committee with some recommendations.

Second, I would just say that in the areas where we do have better control of the border, as I had mentioned, our strategy is based on a deterrence strategy. And when we bring those arrest numbers down, it puts us in a better position to be able to enforce the rule of law. Obviously, we probably make more arrests than any other law enforcement agency in the world, and we recognize fully that the U.S. Attorney's Office cannot prosecute everybody who we arrest. But certainly in those areas where we do have better control of the border, we have excellent cooperation with the U.S. Attorney's Offices, and I think we're making a lot of progress there.

Chairman GRAHAM. Well, please tell your officers we realize how dangerous their job is. It is one of the most dangerous jobs in law enforcement, and we appreciate their sacrifice and service to our country. And that is true of all of you, everyone, because I know this is a very tough business that you are trying to combat.

Senator Cornyn, do you have any further questions?

Senator CORNYN. I just had a couple questions in conclusion. Chief Harris, I know we were talking about the number that Border Patrol catches each year, and I wonder if you would please after the hearing go back and try to provide us an answer, the Border Patrol's best estimate of about how many people come across the border that, for one reason or another, you are unable to catch and they do come into the country illegally each year. Would you get that information for us, please?

Mr. HARRIS. Yes.

Senator CORNYN. And then you mentioned—did I hear you correctly? You said there are now a thousand Border Patrol agents on the 4,000-mile U.S.-Canadian border?

Mr. HARRIS. They're not there yet, Senator. This was just directed. We have gone out with the initial job announcements. We are making selections for those positions now, but we're shooting to have those thousand agents up on the border by the end of this year.

Senator CORNYN. And how many are there now?

Mr. HARRIS. About 600.

Senator CORNYN. Six hundred. And do you know how many agents are on the 2,000-mile U.S.-Mexico border?

Mr. HARRIS. Approximately 9,500.

Senator CORNYN. And I know we have talked to Mr. Hutchinson and others about employing technology, including things like unmanned aerial vehicles. America is the leader in technologies. A lot of the things that we have used to supplement and augment our human resources, I think, are going to be more helpful so that you do not have to literally have an agent posted every few feet on the border. No one is suggesting that that would be a good alternative. But I am hopeful that we will be able to employ the most modern technology we have to be able to assist you in doing what I think we all recognize is a terribly difficult and challenging job.

Thank you. Thank you, Mr. Chairman.

Chairman GRAHAM. I thank the panel. It was very informative. Thank you all for coming very, very much.

Chairman GRAHAM. Thank you very much for coming to our hearing today. We look forward to your testimony.

If you would please rise, I will swear you all in. Raise your right hand, please. Do you solemnly swear the testimony you are about to give the Committee is the whole truth and nothing but the truth, so help you God?

Mr. CHARLTON. I do.

Ms. BOYLE. I do.

Ms. COHN. I do.

Chairman GRAHAM. Again, welcome. Welcome to the Committee. I found the first panel very informative, and I look forward to hearing from your perspective. If you do not mind, we will just do opening statements, and then Senator Cornyn and I will have a dialogue with you.

We will start with you, ma'am.

## STATEMENT OF SHARON B. COHN, SENIOR COUNSEL, INTERNATIONAL JUSTICE MISSION, WASHINGTON, D.C.

Ms. COHN. Thank you, Mr. Chairman, for convening this important hearing on alien smuggling and human trafficking. My name is Sharon Cohn, and I serve as senior counsel and director of anti-trafficking operations for the Institutional Justice Mission, and we believe that the question of meaningful deterrence is the crucial issue in human trafficking. I'm grateful to this Committee for the opportunity to share a little bit about what we've learned at the IJM through its field experience around the world.

IJM deploys criminal investigators in cities around the globe to infiltrate brothels, use surveillance technology to document where victims are located, and then we identify secure police contacts who will conduct raids with us to rescue the victims and arrest the perpetrators. We then coordinate the referral of victims to appropriate after-care and support and monitor the prosecutions.

The IJM investigators have spent literally thousands of hours infiltrating the sex trafficking industry and working with government authorities around the world to bring effective rescue to the victims and accountability to the perpetrators. Through this, I think we have gained some valuable insight as to what provides for meaningful deterrence to brothel keepers and traffickers. I'll limit my remarks to sex trafficking, accordingly.

It's estimated that there are between 18,000 and 50,000 women and girls trafficked into the United States each year. As Federal and local law enforcement agencies and the Justice Department vigorously investigate and prosecute the offenders in the United States, we believe that it is critical that they assist in addressing the issue within a global context.

Like narcotics and arms smuggling, trafficking in persons for sexual exploitation is a multinational crime. Similar to the drug trafficker caught on the borders of the United States, the sex trafficker is part of a series of transactions which includes players in the country of origin, the country of transit, and the country of destination. So in order to effectively disrupt this market, intelligence must be transferred back to both the source and transit country.

And then it is just so vitally important that national and local law enforcement in foreign-source countries have the political will and have the resources to combat trafficking, because our experience has taught us fundamentally this: Sex trafficking reaches the United States because it is tolerated by local law enforcement in countries around the world. In cities around the globe, millions of women and girls are trafficked and offered to customers in brothels. And every day millions of customers around the world find these girls.

In fact, it does not good at all for brothel keepers to keep the victims hidden. To make money on these transactions, they have to hold these victims out to the public not just once but continually day after day for sexual exploitation. Obviously, therefore, if the customers can find the victims, it stands to reason that the police can as well.

How, therefore, do you possibly succeed in committing this crime over and over again, day after day, in front of the open public? You do so only if permitted by local law enforcement, and generally this is facilitated by bringing the police into the business and sharing the profits with them in exchange for protection against the enforcement of laws that are in place and consider all of these a crime.

This truth is most tragically demonstrated through the lives of the little ones that we have had the privilege to assist in rescuing. I want to tell you a little bit about a friend of mine named Simla, who was raised in a village off a main road to a small city in Asia. When she was 14, a woman in the village sold her to a trafficker. She was told she'd be working in a restaurant in the city. When the trafficker brought her to the city, she was sold to a brothel and told she would have to have sex with customers in order to pay off the debt—a debt that wasn't hers but was acquired to her when she was sold by the trafficker to the brothel.

For two and a half years, Simla was subjected to sexual assaults, multiple times a day. She was beaten when she cried, she was beaten when she was sleepy for the customers, and she was beaten when she said she wanted to go home. But the worst beating that she ever received, the one that made it difficult for her to walk, was the beating she received the day after a police officer complained that she didn't smile after he had finished raping her. He would come to the brothel regularly to receive his payment for providing protection for the brothel, and sometimes he would rape the girls instead of receiving cash payment. Simla's friends in the brothel, who were also children, confirmed that other offices regularly visited the brothel and abused the girls.

IJM investigators identified her and others in the brothel as minors and brought this to the attention of local law enforcement. On the night of the raid, a police officer called the brothel keeper and told her that they were coming. The brothel was empty of children by the time we arrived. The children later told us that they had been—they had received a phone call, that they had been loaded on the back of a flat-bed truck, covered with a blanket, and driven across town. Ultimately, senior police officials communicated down the chain of command that the children must be found and released, and Simla and her friends were rescued by the authorities later that day.

In fact, just 2 weeks ago, my colleagues in the field received information concerning a case where the police are not only providing protection for the brothel but, in fact, were employed by the brothel to find and return two girls who had escaped from the brothel. The police returned to the two girls, beaten, and these girls were subsequently shot and killed by the brothel keeper.

Stories like these are repeated throughout the world where local law enforcement do the bidding of traffickers and brothel keepers. Without police protection, the brothel keeper cannot succeed, and with it, he simply cannot fail. Once the police switch sides, the brothel is fatally vulnerable and effective law enforcement can provide rescue and secure arrests. Until they do, it is the girls that are fatally vulnerable.

So, in the end, the brothel keepers only care about two government actions: Is the government seriously threatening to actually put me in jail for this crime? And is the government seriously threatening to remove the police protection that I have paid for?

What can the U.S. Government do to create a meaningful deterrent for sex traffickers? On the domestic front, we believe the U.S. Government should fully utilize the Federal witness protection program to provide resources to adequately protect family members of cooperating victims who continue to live at risk in source countries. All law enforcement depends upon the cooperation of victims, and the United States has the opportunity to set the standard for the world by the way we treat the women and girls who are trafficked into this country. By employing the T visas and providing witness protection, authorities can create a safe environment for cooperation.

Second, we believe the United States must aggressively prosecute sex tourists who create the demand for trafficking victims. We commend Congress for passing the PROTECT Act, which frees the U.S. Attorney's Offices to vigorously prosecute sex crimes committed by Americans abroad. We must encourage prosecution and conviction of these criminals and the subsequent media coverage to deter other would-be criminals.

We had a recent experience in Cambodia that is more fully explained in my written testimony, but, in short, I will tell you the multinational aspect of this crime and American culpability is well displayed in this case, where little girls as young as 5 years old are trafficked from Vietnam into Cambodia. American sex tourists, who learn about this place from the Internet and from Internet chat rooms and websites, travel to Cambodia specifically for the purpose of sexually exploiting these little girls.

We worked with the Cambodian authorities and with Ambassador Charles Ray in order to secure the rescue of 37 girls, the youngest of whom was 5 years old, and the arrest of 16 perpetrators who are now charged under Cambodian law and are awaiting trial.

In addition, U.S. Customs is following leads resulting from the raid, and we are hopeful that an investigation will result in convictions of identified American sex tourists.

It's hard for me to describe to you, Senators, how horrible it is to sit and talk to these girls and have them describe to me how it was the Americans who made them sleep with them, it was the Americans who they had to spend time in bed with, it's the Americans who they were forced to service, and they were beaten in order to do so. It is simply imperative that the United States crush the demand that's created by its own citizenry.

Third, we would encourage Federal law enforcement agencies to continue to communicate through joint training initiatives and funding that sex trafficking is a priority issue for the United States and that it's a violent crime worthy of attention of elite law enforcement.

Law enforcement priorities are set by senior-level political authorities. Where the United States encourages engagement on a senior level, it will be able to influence the priority and conduct of street-level enforcement. Like counter-narcotic initiatives, the United States should commit to invest significant financial and personnel resources to ramp up and improve counter-trafficking initiatives overseas.

And, finally, we would just encourage the United States to improve its information-sharing mechanisms with foreign-source countries so that law enforcement attaches in the United States embassies overseas will receive the information that is obtained through thorough debriefings of victims and suspects here in the United States and be able to convey that to local law enforcement overseas.

We just believe the United States has the capacity to create a meaningful deterrence, not limited to its own borders, through proactive engagement with law enforcement in foreign countries, and we are very appreciative for the Subcommittee for calling this hearing.

Thank you very much.

[The prepared statement of Ms. Cohn appears as a submission for the record.]

Chairman GRAHAM. Thank you very, very much. We have a lot of questions for you, I think.

Ms. Boyle, welcome.

## STATEMENT OF JANE J. BOYLE, UNITED STATES ATTORNEY, NORTHERN DISTRICT OF TEXAS, DALLAS, TEXAS

Ms. BOYLE. Thank you. Chairman Graham, Senator Cornyn, I am Jane Boyle, the United States Attorney for the Northern District of Texas. I will tell you that it is an honor to appear here before you today to recount a recent case in my district that illustrates the tragedy of the human trafficking form of alien smuggling and I also think typifies the difficulties that are posed to prosecutors in prosecuting cases of this nature.

Let me go back to the beginning. In early April 2002, the FBI and former INS, along with many factions of local law enforcement, initiated a joint investigation into what we dubbed "the Molina organization." Through the investigation, we learned that between December of 1998 and May of 2002, the organization smuggled approximately 200 impoverished young Honduran females into the United States. Some of these young ladies were as young as 14, we think perhaps even as young as 13.

We also were able to determine from the investigation that the Molinas enticed these young women, and their families to allow them to go, into going by promising them a better life in the United States, work as housekeepers and nannies and waitresses. To prevail on their impoverished parents to let them take the trip and make the change in life, they did make this promise to the parents, but they also required that the parents, in order for this benefit they promised the girls were going to get, pay a large sum, which caused many of them to give up the deeds to their property in order to finance the trip.

By all of the girls' accounts, Senators, the trip from Honduras to Fort Worth was horrific for these young ladies. They described firsthand spending weeks traveling through mountains and deserts, walking at night to avoid detection, and often going several days without even any food or water.

To enter the United States from Mexico, some of them were located into wooden compartments installed under trucks. They were packed in these compartments, some of them, head to toe for as long as 10 hours. Many, as you might expect, even urinated on themselves in the process, and many suffered injuries.

Upon arrival in Fort Worth, what they thought was the promised land of jobs as waitresses turned into them immediately being clothed in risque outfits, and this was immediate, and then compelled to work in one of four Fort Worth area bars controlled by the Molinas. Some of the young women were coerced into prostitution, and others were required to what they, I would say, loosely described as "dance" with customers, the male customers, to encourage the sale of overpriced drinks. And we're talking about $10 beers, I think is what the facts were.

They earned $100 for a mandatory 60-hour work week, all of which was applied to their exorbitant smuggling debt. They earned nothing at all if they didn't meet the 200 per week quota of selling drinks.

Undercover agents—and these are veteran FBI agents and former INS agents—who saw the young women in the bars were struck and moved by the frightened and upset faces that characterized these little girls. Some of them described them as huddled together, once they go there and realized what it is they were going to be doing for a living, huddled together holding hands in the middle of the bar. And the agents were as moved as I've ever seen Federal agents working undercover.

The women were forced through intimidation to live in residences during this time under the Molinas' control, and some of their debts were as high as $10,000 and $12,000 that they had to repay before they were allowed to leave. Many were verbally abused and intimidated with threats of capture by immigration authorities. And, again, the Molinas also threatened to take the victims' families' properties—you can imagine these were not lavish properties, what they had—from their families in Honduras if they didn't pay off their debts.

We determined that the smuggling operation the Molinas controlled was very lucrative and quite sophisticated. Without getting into great detail, we found that during the time period that we watched the investigation, approximately $1.7 million was wire transferred from locations throughout the United States to Molina smugglers throughout Latin America and in North America.

The investigation culminated in the execution of search warrants and administrative inspections of several bars and restaurants in Fort Worth. In all, we detained 93 individuals out of the arrests and search warrants; 34 of those we determined to be actually the smuggling victims.

Thirteen defendants, three of which are now fugitives, were eventually charged with various violations of the Federal immigration laws, including smuggling illegal aliens into the United States, our familiar Section 1324 of Title 18 statute. Five of the defendants were also charged under the TVPA, which I will say as an aside has been a wonderful tool for prosecutors since it's been enacted. Ten members of the organization that we were able to capture and charge pled guilty to the smuggling charges. Their sentences were around 5 years, some over, most under 5 years in prison.

I want to, if I could for just a minute, detail the difficulties the prosecution team, Richard Roper and Rose Romero, faced during the course of the case, if I could have just a minute, Senators.

Several of the victims' families were threatened by Honduras— were threatened in Honduras by fugitive defendants and their accomplices. The Molinas threatened to burn the victims' family members' houses or even kill them if their daughters testified once we were arrested and charged the individuals. Unfortunately, we could not guarantee the safety of the families in Honduras, and as a result, many of the young women were intimidated and reluctant to testify against their traffickers.

The second barrier that we faced, which is characteristic of cases of this nature, is we faced a formidable cultural barrier with respect to the forced prostitution charges. Most of these young ladies came from strict, conservative, Catholic homes and refused to publicly admit that they'd been forced to engage in nefarious activities such as prostitution. And that stymied our ability to charge them under the TVPA as we would have preferred, all of the defendants.

More problems arose when three of the defendants and several material witnesses fled to Honduras, and we determined that the extradition treaty between the United States and Honduras prohibited their extradition.

Despite these difficulties—and our efforts to rescue the victims I believe were successful, Senators—nearly all of the 34 trafficking victims we have been able to, through the great benefit of the TVPA, involve them in what we've called the continued presence program, which is a benefit Congress has given us under the TVPA. They are receiving assistance and hopefully working down the road towards a visa in the United States. But, again, that's through the benefit of the TVPA.

I thank you for your time and attention and your indulgence, and I appreciate the opportunity to speak and welcome any questions.

[The prepared statement of Ms. Boyle appears as a submission for the record.]

Chairman GRAHAM. Mr. Charlton?

## STATEMENT OF PAUL K. CHARLTON, UNITED STATES ATTORNEY, DISTRICT OF ARIZONA, PHOENIX, ARIZONA

Mr. CHARLTON. Chairman Graham, Senator Cornyn, good morning. My name is Paul Charlton. I am the United States Attorney from the District of Arizona. I'd like to begin by telling you what a distinct honor and a privilege it is to be here today to share with you some of the cases and some of the events that have taken place in the District of Arizona as they relate to alien smuggling. These cases raise concerns about the appropriateness of the penalties for alien smuggling under the current Sentencing Guidelines, particularly in cases that involve the risk of serious injury, death, or that actually result in injury and death. It's my understanding, Senators, that the Sentencing Commission is currently looking towards adjusting some of those sentencing guidelines, and I look forward to the Department of Justice, the Congress, and the Sentencing Commission addressing this very important issue and am thankful to you and your leadership for bringing this issue to our attention as well and giving us the opportunity to discuss this with you.

The District of Arizona faces especially daunting challenges in combating alien smuggling. The risks inherent in transporting human beings through the harsh and unforgiving desert of southern Arizona, as well as the increasing violence in smuggling, has resulted in a disturbing humanitarian crisis.

In my written statements, Senators, I've provided a number of case examples that show the disparity between sentences that an individual can receive in human trafficking cases and human smuggling cases. But, Senator, because your questions and yours, Senator, have focused on the idea of an inequity especially in human smuggling cases as opposed to human trafficking cases, I will, with your permission, address my comments here now to those cases and the disparities in sentences that relate only to human smuggling cases.

In the case of United States v. Miguel and Johnson, the prosecuting attorney, Sara Tessakai, in the United States Attorney's Office charged the defendants with three counts of alien smuggling involving risk of death where they had transported three children in the trunk of a vehicle during a hot July afternoon in Tucson, Arizona. One of the minors was found unconscious, unresponsive, and had to be revived by medical personnel called out to the scene. The defendants pled guilty to the indictment without benefit of a plea agreement, and yet they receives sentences of only 21 and 37 months, respectively.

In another cases, United States v. Alderette-Moreno and Loera-Chavez, two defendants smuggling 19 illegal aliens instructed the aliens to get into a van, where they were required to lie on the floor of the vehicle and literally pile one on top of the other for lack of room. And the van door was broken and did not close, so that one of the aliens had to hold that door closed while the vehicle was in motion. When one of the van's tires blew, the vehicle rolled, killing the man who was holding the door and permanently paralyzing another alien from the neck down and seriously injuring other occupants of the van. Again, the defendants pled guilty, without benefit of a plea agreement, to conspiracy to transport illegal aliens and alien smuggling where death resulted. One defendant was sentenced to 48 months, the other defendant to 57 months' incarceration.

Finally, in United States v. Diego Gallegos Castillo, after walking 4 to 8 hours in the southern deserts of Arizona near the Tohono O'Odham Indian Reservation, and remaining in the wash for a full day and a night, a large group of aliens was picked up by smugglers who instructed them to pile into a pickup truck. Approximately 11 aliens piled into the bed of the truck while three entered the extended cab, which did not have a seat, and where a 14-year-old was required to lay across the laps of a number of individuals. While the vehicle was traveling at approximately 87 to 93 miles per hour, the truck flipped over into a wash and resulted in the deaths of four aliens. Serious injuries were sustained by the surviving aliens. This case went to trial, Senators, and the defendants were found guilty of 11 counts of transporting illegal aliens resulting in death or serious bodily injury. The defendant was sentenced to 72 months' incarceration.

I hope the cases I have presented for your review have been helpful for evaluation of the current state of the Sentencing Guidelines with respect to alien smuggling cases.

Thank you, Senators, for your time and attention to this very important issue.

[The prepared statement of Mr. Charlton appears as a submission for the record.]

Chairman GRAHAM. Senator Cornyn?

Senator CORNYN. Thank you, Mr. Chairman.

I would like to first direct questions to Mr. Charlton and Ms. Boyle about cooperative efforts with local and State law enforcement authorities. I think we have heard from the first panel that the Federal Government probably cannot do all this by itself, including those of you who are the front-line prosecutors for Federal law violations.

But, Ms. Boyle, would you perhaps, in the Molina case that you described, talk about the role, if any, of local and State law enforcement authorities to complement the work that you did, and if there was not an involvement of those law enforcement officials, whether you think that would be beneficial or whether you would consider that to be problematic for some reason?

Ms. BOYLE. Senator, let me tell you first generally that State, local, and Federal authorities have never been—have never worked as well together as they have since September the 11th on all fronts.

In this particular case, to hone in on that, I don't know that we would have gotten the investigation off dead center if we hadn't had tips that were initially given in 2001 to the Fort Worth Police Department. I believe they received some anonymous letters. They were extremely cooperative, and this is the Dallas Police Department, the Fort Worth Police Department in particular, and the sheriff's offices of both cities, were very, very involved in our efforts in the Molina case with the former INS and the FBI.

We could not have done this without the efforts of those local agencies. They helped us plan the operation. Probably the most crucial part of this operation was what we call the raid, and that was the evening that we went in under very precarious circumstances. You're going to bars at night with all sorts of individuals frequenting them. So under very precarious circumstances, we planned this together with the local law enforcement authorities, and it turned out very successful. But we could not have done it without their help, without their intelligence. Their intelligence was very key. We are finding these days that local law enforcement play a crucial role in providing us the intelligence we need to carry out major Federal operations. So this was a perfect example of that.

So, from my experience in the Molina case, it's a prime example of the great work of local law enforcement and their contribution to our effort.

Senator CORNYN. Mr. Charlton, what about your experience?

Mr. CHARLTON. Senator, in April of this year, we initiated in our district Operation Desert Risk, which was our attempt to address the very issue that you are talking about, Senator. We invited State, local and Tribal law enforcement officials to join Federal law enforcement officials in addressing the issue of alien smuggling on the border.

We received a very good response from State, local and Tribal law enforcement officials because everyone recognizes that alien smuggling is not just a Federal problem or issue. It is a problem which affects all of our local communities as well and begins to affect the quality of life issues for our State, local and Tribal law enforcement officers and the community.

We are continuing this operation through the summer months, which are our most critical months in Arizona, and we will terminate this operation in October, after which we will look at the data which we have collected to determine whether or not we have, in fact, seen an effective reduction in the kinds of crimes, both in smuggling and in collateral offenses that surround smuggling offenses in the District of Arizona.

Senator CORNYN. As Ms. Boyle knows, in Texas, when I was in a previous life as attorney general of Texas, we worked closely, as State law enforcement officials, with local and Federal authorities on gun crime, and it was a—what we encountered, before we initiated the Texas exile program is what, frankly, I think what the 9/11 report demonstrated, Mr. Chairman, at the Federal level is we were not playing as well together as we should or working as well together as we should, and I know I am gratified to hear what both of you have said about the work that you are doing, but we just have to figure a way to get more resources into the game, and particularly when we are talking about homeland security, we are talking about these horrific crimes that you described, I just think we need to do everything we can in the Congress to encourage and, indeed, to require that sort of cooperative effort.

One area that I know has been, it does not necessarily relate or it might, to some extent, to what we have been talking about here today, but it is simple information sharing. And you mentioned the intelligence, Ms. Boyle, that you received from local law enforcement officials. Well, the problem is, I am afraid, too many times it comes from the bottom up, but it does not come from the top down. In other words, Federal intelligence agencies do not share information not only with one another, but they do not share it with State and local law enforcement officials, and I know there are concerns with regard to maintaining the confidentiality of that information, but there are certainly methods to eliminate the most sensitive information and provide the functional information that is needed in order to fulfill the most complete role possible when it comes to dealing with crimes of this nature and other nature as well.

Ms. Boyle, in terms of the organization of the Molina organization, would you describe for us, I mean, are we talking about sort of the mules, I mean, the low-level people that you were able to successfully prosecute. You said some remained in Honduras and otherwise. Were you able to cut the head off the snake or were you dealing with some lower-level people or some combination?

Ms. BOYLE. We did get four individuals that we believe were at the top of the organization, Senator, as well as the mules. Again, thanks to all of the intelligence that we were able to gather. It is a very lucrative business, and it is operating out of a very poor country, so the incentive is almost impossible to stop someone's inclination towards getting involved in that.

I believe this organization was fairly well bashed. I will tell you, though, that that does not mean others will not spring up in its place, but I think we just have to be ready to go after them when they do and make sure they know that the United States will not tolerate this kind of crime.

But we were able to get individuals at the top of the organization, I believe at the very top, and for reasons I can easily and would be glad to get into, however, we were not able to secure the types of sentences I think that these individuals deserved.

Senator CORNYN. You said they were 5 years, on the order of 5 years?

Ms. BOYLE. They were. We had one over five and I think three others that were slightly under five.

Senator CORNYN. I agree with you that, given the nature of the offense, that that does not seem like a proportional sentence, and we would be interested in learning more about your suggestions as to what we can do to address that.

Ms. Cohn, let me, as the father of two daughters, I must tell you, and just as a human being, the story that you conveyed about sex trafficking and what we might be able to do about it is chilling. I guess my question to you, though, is if there are Americans who are committing these offenses in foreign countries, what sorts of things specifically can we do, this Congress do, recognizing that things that happen in foreign countries are difficult for our laws to reach, what sorts of things can we do, in your opinion, to get at and address those sorts of heinous acts?

Ms. COHN. Senator Cornyn, the Congress has graciously and brilliantly, I think, passed the Protect Act, which changes the sex tourism law that was currently on the books. The old law required that the U.S. Attorney's Office be able to demonstrate that a sex tourist traveled overseas for the purpose of exploiting, committing a sexual crime, but they had to prove that the intent was formed in the United States in order to secure the jurisdictional hook.

The new Protect Act takes that away and simply requires that the U.S. Attorney's Office prove that an individual traveled overseas and committed an act, and removes that intent requirement domestically. The old requirement sort of suggested that the FBI investigators would be able to find neighbors of the sex tourists in Oklahoma or somewhere and have them say, yes, he told me he was going overseas to have sex with a 6-year-old—very difficult to prove.

Now, that is gone, and all you have to prove is that the person did, in fact, travel to Cambodia, have sex that was illegal in the United States, for example, sex with a minor, and you can prosecute them here in the United States. And these are sex tourists who travel and come back. They are businessmen. They are doctors or lawyers. They travel, spend a lot of money to commit a crime they do not believe they could get away with committing in the United States, so they travel overseas to commit that crime, and now I think Congress has given the Department of Justice the tools that it needs to convict them here in the United States.

I should also add that it really, in my opinion, only takes a few meaningful convictions to create that deterrent effect that you were talking about because these areas get known, for example, this area of Svay Pak in Cambodia, become known because of websites and Internet chat rooms.

The example of how effective the deterrents can be is after we were able to secure the raid in Cambodia and rescue the girls and have the perpetrators arrested, the Internet chat rooms that cover that area called Svay Pak were saying the party is over in Cambodia. And that is the kind of information flow that you want to generate, and I think a few convictions here in the United States would send that message that sex tourism by Americans is not going to be permitted any more.

Senator CORNYN. Thank you, Mr. Chairman.

Chairman GRAHAM. Thank you, Senator Cornyn.

For the record, we have a statement here from Senator Biden that I would like to introduce into the record and Senator Leahy, and we will hold open the ability of members to make written questions to the Subcommittee and to the witnesses for one week.

I would just like to acknowledge Senator Cornyn's contribution to making this hearing happen, and he has been a very good friend, a valuable ally and tried to bring some attention to this. So thank you very much, Senator Cornyn.

Ms. Cohn, we will start with you.

I am familiar with the Protect Act and what we are trying to accomplish, but it seems to me that the real gap is foreign countries, and their willingness to allow this to continue. Is there anything that you can think of, other than the Federal Witness Protection Program, where we can have people more forthcoming about prosecutions that we can do to bring more attention, and what countries are you talking about?

Ms. COHN. Mr. Chairman, we work throughout the world, and I have spent a lot of time on sex trafficking in South Asia and Southeast Asia and would draw attention to those ares in particular, but I think that the TVPA that Attorney Boyle was discussing is one of those instruments that Congress has provided the U.S. Government to help combat trafficking by addressing foreign countries.

The TVPA requires that the State Department rank countries' performance in their efforts to combat sex trafficking, whether they are meeting minimum standards or whether they are making significant efforts to meet those standards to combat trafficking.

I think that tier-rating system does help. I would be excited to see it employed more vigorously, particularly as the State Department examines how complicit Government officials are in those countries and whether those countries can provide evidence, documentation that they have disciplined corrupt police officials and that they have convicted brothel keepers and traffickers.

Mr. Chairman, I would just sort of direct you to the TIP report, the Trafficking In Persons report, that just came out in June. And if you look through it, it is required to discuss the prosecutorial effectiveness in certain countries as they go against the traffickers and brothel keepers. And you will find in a great majority of countries, that prosecutions are abysmal, to an extent, difficult to articulate today, but that it is difficult, almost impossible, to secure a conviction in those countries. And I think that the U.S. Government, including this Congress, can, in all of its communications with those countries, demonstrate how serious the U.S. Government takes this issue and that failure to demonstrate that significant efforts are being taken in those countries to combat trafficking will lead to penalties and sanctions by the U.S.

Chairman GRAHAM. If you could provide sort of your view of this to supplement the June report about what you think are the most egregious countries, in terms of lack of prosecution, I would appreciate it.

What are the numbers that we are talking about of Americans who have been convicted for going overseas in sex tourism?

Ms. COHN. I think the conviction numbers for Americans is dramatically low. I don't have the numbers—I don't have the numbers available, but—

Chairman GRAHAM. Any idea of the volume of how big a business this is?

Ms. COHN. I think I read a statistic that ECPAT, which is the End Child Prosecution and Trafficking NGO organization, I think estimates that 25 percent of sex tourists are Americans.

Chairman GRAHAM. Do you have a number on that. Are we talking thousands of people?

Ms. COHN. You are certainly talking thousands of people. I am not sure if you are talking more than that.

Chairman GRAHAM. Thank you very much.

Ms. Boyle, the Fort Worth case I think illustrates some success and a problem, and that is the Honduran Government. I am very concerned about foreign countries' willingness to help us address a global problem. At the end of the day, what kind of cooperation did you get from Honduras, if any?

Ms. BOYLE. We did get some cooperation, Mr. Chairman. They did help us. At one point, some of our prosecutors, they did help us, Mr. Chairman, not to the extent we would have liked, but I will tell you they did not cut us off.

They, at one point, some prosecutors and agents traveled down there to find out a little bit more about the threats to the families and that status, and they were able to at least witness in one situation an actual threat to a family member, and they were able to secure from local police what we call up here a protective order of that nature. I am not sure how effective it was outside of just the paper it was on, but they were helpful.

I think the treaty hurt us without—because we could not get the extradition. I can't compare them to other countries, but I have talked to the agents and the prosecutors before I left to ensure that I was aware of exactly where they stood on the Honduran Government's cooperation, and they felt that they were cooperative. Their hands were somewhat tied by their own laws and really their lack of knowledge about the worldwide focus on this.

But I also think, and I agree so much with Ms. Cohn on this very, very important provision of the TVPA, with the tiering of the countries, as long as that is focused on and enforced vigorously, I think that is going to be a big help to our Government to get other Governments to cooperate.

Chairman GRAHAM. Ms. Cohen, ASEAN is a group of Asian Nations. Is there any counterpart to the Protect Act or the TVPA Act in Asia or Europe?

Ms. COHN. Mr. Chairman, most all of the countries where we work have laws that prohibit the acts that we are talking about, in one manner or another, whether it is just prohibiting child rape or it is prohibiting kidnapping or prohibiting trafficking for fraud or more explicit or less explicit. The problem is actually not, generally not in what the statutory provisions are, as it might be in the U.S., that that constrains the U.S. attorneys, but rather that they are simply not enforced.

Chairman GRAHAM. Mr. Charlton, now, you mentioned some sentences that seem very low, given the activity involved. When the Sentencing Guideline Commission meets, will you have input as to how that should be changed, your organization?

Mr. CHARLTON. Mr. Chairman, I am sure that through the Department of Justice we will have an opportunity to make our views and our perspectives known.

Chairman GRAHAM. I am very interested in the Committee getting involved also in working with the Sentencing Guideline Commission. I will make the same offer I made to the last panel. We have statutes in place now that we did not have before. They seem to be working. They seem to empower you to do a better job.

What I would like is sort of make your shopping list. Given that dynamic, what can we do to enhance the viability of these statutes and what things are left undone in terms of the law and resources and provide that input to the Committee, and we will try to meet your needs the very best we can.

I understand, Ms. Cohn, you have some photos or some pictures.

Ms. COHN. I have one photo just of Simla to help you appreciate her age. I did not bring photos of the small Cambodian, the Vietnamese children in Cambodia. She is currently in aftercare in Southeast Asia, and I just visited her not long ago. She is doing quite well. I would like to bring her to the States, at some point, but she is one of millions of girls who, through no fault or decision-making of their own, are simply taken advantage of by the greed of others, and law enforcement participates in that.

And I should add only, because I have such great concern for her, that if the police are not directly complicit in actually killing these girls, as they are in the case where they are returned to the brothel keeper, and the brothel keeper shot them, they are complicit in the deaths of these girls to the extent that the HIV/AIDS transmission among sex trafficking victims is just so brutally and extraordinarily high, that to be trafficked into that enterprise is essentially a death sentence for young girls.

Chairman GRAHAM. What country are we talking about where she was involved?

Ms. COHN. Simla was trafficked in Thailand.

Chairman GRAHAM. Well, thank you all very, very much. God bless you in your efforts to deal with this problem. You have our full support and encouragement. I am sure that is bipartisan in nature, and please take us up on the offer to strengthen the current laws and to give you more resources.

And for lack of a better word, we are just dealing with scum here, and we need to act in a cohesive manner to make the world a better place.

God bless. Thank you for coming.

The hearing is adjourned.

[Whereupon, at 11:38 a.m., the Subcommittee was adjourned.]

[Submissions for the record follow.]

[Additional material is being retained in the Subcommittee files.]

SUBMISSIONS FOR THE RECORD

**"Alien Smuggling/Human Trafficking:
Sending a Meaningful Message of Deterrence"**

**Subcommittee on Crime, Corrections and Victims' Rights
Committee on the Judiciary**
Friday, July 25, 2003
Dirksen Senate Office Building at 10:00 a.m.

Written Statement of Joseph R. Biden, Jr.

I want to thank the Chairman Graham for convening today's hearing on alien smuggling and human trafficking. Unfortunately, this is one of the fastest growing areas of international criminal activity and one which requires a coordinated and aggressive response by the U.S. government.

Alien smuggling and human trafficking are two crimes that we cannot ignore either here at home and abroad. Similar to the illegal drug markets, trafficking and alien smuggling are global crimes with entrenched roots. The numbers are extraordinary. Between 700,000 and 4 million people are believed to be trafficked each year worldwide – the majority of them are women and children. A recent report from the State Department estimates that more than 50,000 women and children are trafficked to the United States. Make no mistake, smuggling and trafficking are big business. Human smuggling has become an international lucrative criminal market -- like narcotics and weapons trafficking -- that generates an estimated $9.5 billion per year.

I was honored to be a member of the Conference Committee which authored the "Trafficking Victims Protection Act of 2000," which the Congress overwhelmingly passed. And I look forward today to hearing from our witnesses about that important Act's implementation. I want to make sure that we have provided law enforcement all of the legal tools and financial resources they need to go after criminals who smuggle aliens or engage in sexual trafficking. I want to find out if criminal penalties – both here and abroad – are sufficient to deter traffickers. I want to explore if there are innovative things which can be done with extradition and witness protection to encourage fearful victims and witnesses to come forward to help make these cases.

But I also want to learn from our witnesses whether we are "connecting the dots" between domestic prosecution here at home, and what is happening abroad, in the source countries which supply the tens of thousands of people smuggled or trafficked into this country each year. It's not enough to just prosecute U.S.-based traffickers. That would be like aggressive domestic-based drug prosecution which ignores coca cultivation in Colombia or poppy production in Afghanistan. Our successes to date in the war on terrorism demonstrate our ability to secure the cooperation of foreign governments in combating criminal activity abroad. We need to make sure we seek a similar level of cooperation from foreign governments which supply the very humans smuggled and trafficked into this country and others around the world.

Mr. Chairman, I again commend you for shining a spotlight on this important issue. We have made some good progress in the last two years and I look forward to hearing how we can do even more.

# Department of Justice

STATEMENT

OF

JANE J. BOYLE
UNITED STATES ATTORNEY
NORTHERN DISTRICT OF TEXAS

BEFORE THE

SUBCOMMITTEE ON CRIME, CORRECTIONS, AND VICTIMS RIGHTS
COMMITTEE ON THE JUDICIARY
UNITED STATES SENATE

CONCERNING

HUMAN TRAFFICKING AND ILLEGAL IMMIGRANT SMUGGLING

PRESENTED ON

JULY 25, 2003

**STATEMENT OF UNITED STATES ATTORNEY JANE J. BOYLE**
Northern District of Texas
Before the Crime, Corrections and Victims Rights Subcommittee
Senate Judiciary Committee
July 25, 2003

Chairman Graham, Ranking Member Biden, and Members of the Subcommittee, I am
Jane Boyle, the United States Attorney for the Northern District of Texas. It is an honor
to have the opportunity to appear before you today to recount a recent case in my District
which illustrates the tragedy of the human trafficking form of alien smuggling and which
typifies the difficulties posed in prosecuting a case of this nature.

Although the Northern District of Texas is not on the Texas/Mexico border, two
interstate highways which run through the District are prime transportation routes for
smugglers. Last year, successful law enforcement efforts led to the disruption of a major
human trafficking operation out of Honduras. This operation, we dubbed the "Molina
Organization," involved the trafficking of approximately 200 young women between the
ages of fourteen and thirty-five to Fort Worth, Texas. Permit me to brief the
subcommittee on that investigation and share with you some of the issues and obstacles
we faced in prosecuting that case.

In late 2001, Fort Worth police received anonymous letters complaining that young
Honduran women were being smuggled into the United States and forced to work as
prostitutes in bars in the Fort Worth area. In early April 2002, the FBI and former INS,
along with the Fort Worth and Dallas police departments, and other law enforcement
agencies, initiated a joint investigation into the Molina Organization. Through the
investigation we learned that between December 1998 and May 2002, this organization
smuggled approximately 200 impoverished young Honduran females, some as young as
fourteen, into the United States. We also determined that the Molinas enticed the women
into the operation by promising them employment as housekeepers or waitresses in
restaurants in Fort Worth, in exchange for an undisclosed "fee." To prevail on their
impoverished parents, the Molinas promised them that the young women would enjoy a
better life in the United States. To finance the trips, the Molinas required many of these
families to give up the deeds to their homes and properties as collateral for smuggling
debts.

By all accounts, the trip from Honduras to Fort Worth was horrific for these women.
They described spending weeks traveling through mountains and deserts, walking at night
to avoid detection and often going several days without food or water. To enter the
United States from Mexico, some were loaded into wooden compartments that were
installed under trucks. They described being packed in hidden compartments "head to

toe," for 10 hours without food or water.  Many urinated on themselves and several suffered injuries.

Upon arrival in Fort Worth, the young women were immediately clothed in risque outfits and then compelled to work in one of four Fort Worth-area bars operated by the Molinas. Some of the young women were coerced into prostitution, and others were required to "dance" with male customers to "encourage" the sale of overpriced drinks.  They earned $100 for a mandatory 60 hour work week, all of which was applied to their debt, and they earned nothing if they failed to meet the $200 per week drink quota.  Our surveillance revealed that forty to fifty young Honduran women were driven daily to the four bars. Undercover agents who saw the young women at the bars described some of them as very frightened and upset.  The women were forced through intimidation to live in residences under the Molinas' control until their exorbitant smuggling debts, often as high as $10,000 , were paid in full.  They were instructed to hide from law enforcement authorities and, if asked, to lie about the nature of their employment.  Many  were verbally abused and intimidated with threats of capture by immigration authorities.  The Molinas also threatened to take the victims' families' properties in Honduras if any of the women left before their debt was paid.  In total, we estimate that the Molina's smuggled over 200 Honduran women into the United States.

The smuggling operation was sophisticated and quite lucrative.  The Molinas solicited the services of various individuals in San Pedro Sula, Honduras, Esquipula, Guatemala, and Los Angeles, California, as well as other locations in Central America, Mexico, and the United States, to smuggle the victims from the Republic of Honduras to Fort Worth, Texas.  The Molinas paid the smugglers by wire transferring money through various financial institutions.  Between February 1999 and  May 2002, the Molinas wire-transferred in excess of $250,000, through Western Union alone, from Fort Worth to various locations along the smuggling route.  In addition, we found that during this time period,  approximately 1.7 million dollars was wire transferred from locations throughout the United States to Molina smugglers in Esquipula, Guatemala, Mexico, Los Angeles, and South Texas.  The Molinas also accumulated a considerable amount of personal wealth through their smuggling operation.  Honduran officials reported significant cash deposits in Molina bank accounts and numerous property purchases.

The investigation culminated in the execution of search warrants and administrative inspections of six bars and six residences in Fort Worth.  Approximately eighty individuals were detained on immigration violations.  Twenty-five of the eighty were identified as trafficking victims. Subsequent searches, executed on three residences and a bar, yielded thirteen additional detainees, nine of whom were trafficking victims. Thirteen defendants were eventually charged with various violations of the federal immigration laws, including smuggling illegal aliens into the United States in violation of

Statement of United States Attorney Jane J. Boyle - Page 2

Title 8, United States Code, Section 1324.  Five of these defendants were also charged with obtaining labor and services by threats of serious harm and physical restraint, in violation of Title 18, United States Code, Section 1589.  Three defendants remain fugitives.  Two of these were leaders of the Molina Organization and fled to Honduras days after the warrants were executed.  Thirty-four victims qualified for the "continued presence program" under the Trafficking Victims Protection Act of 2000 and were placed with private relief agencies.

Prior to trial, ten members of the organization pled guilty to the smuggling conspiracy or related charges.  The four top members of the organization were sentenced to five years imprisonment.  The others received sentences below five years.

The prosecution team faced numerous difficulties during the course of the case, a few of which I will highlight for the subcommittee.  First,  several of the victims' family members were threatened in Honduras by fugitive defendants and their accomplices.  The Molinas threatened to burn victims' family members' houses or even kill them if their daughters testified.  Unfortunately, we could not guarantee the safety of these families in Honduras, and as a result, the women were intimidated and reluctant to testify against their traffickers..  Secondly, the prosecutors faced a formidable cultural barrier with respect to the forced-prostitution activities of the defendants.  Many of our victims, who came from strict, conservative, Catholic homes, refused to publicly admit to engaging in prostitution, a fact that stymied our ability to charge the defendants with more serious crimes, such as, Mann Act and related violations.  More problems arose when three defendants and several material witnesses fled to Honduras, and we determined that the extradition treaty between the United States and Honduras prohibited the extradition of Honduran nationals to the United States.  We also found that the treaty  provided no practical mechanism to obtain the testimony of unwilling, Honduran material witnesses.

In addition, prosecution under the forced labor statute,  Title 18, United States Code, Section 1589, proved problematic.  The statute fails to actually define the terms "physical restraint" and "serious harm" as a means to obtain labor.  Although the victims in this case were required to live and work as dictated by the Molinas, the evidence did not clearly establish that they were held by actual physical force or violence.  Moreover, this statute does not criminalize obtaining the labor of these women by fraud and deception, one of the major components of this case.

Despite these difficulties, our efforts to rescue the victims and disrupt this smuggling operation were successful.  Nearly all of the thirty-nine trafficking victims we were able to identify as a result of this investigation were placed in the "continued presence program" have done  well in this country and are working hard to achieve better lives.

Thank you for your time and attention.  I appreciate the opportunity to speak on this important and timely matter.  I would be pleased to answer any questions the members might have.

# Department of Justice

STATEMENT

OF

**PAUL K. CHARLTON**
**UNITED STATES ATTORNEY**
**DISTRICT OF ARIZONA**

BEFORE THE

**SUBCOMMITTEE ON CRIME, CORRECTIONS, AND VICTIMS RIGHTS**
**COMMITTEE ON THE JUDICIARY**
**UNITED STATES SENATE**

CONCERNING

**HUMAN TRAFFICKING AND ILLEGAL IMMIGRANT SMUGGLING**

PRESENTED ON

**JULY 25, 2003**

**STATEMENT OF UNITED STATES ATTORNEY PAUL K. CHARLTON**
District of Arizona
Before the Crime, Corrections and Victims' Rights Subcommittee
Senate Judiciary Committee
July 25, 2003

Concerning

HUMAN TRAFFICKING AND ILLEGAL IMMIGRANT SMUGGLING

Chairman Graham, Ranking Member Biden, and Members of the Subcommittee, I am Paul Charlton, the United States Attorney for the District of Arizona. I am honored to have the opportunity to appear before you today to recount recent cases in my District which illustrate the human tragedy of alien smuggling and which highlight the difficulties involved in prosecuting cases of this nature. These cases also raise concerns about the appropriateness of the penalties for alien smuggling under the current Sentencing Guidelines, particularly in cases that involve the risk of serious injury, death, or that actually result in injury or death. I understand that the Sentencing Commission is considering amending these penalties. I hope that the Department, the Commission, and the Congress can work together to make such that such crimes result in appropriate prison terms that reflect both the inherent risk of border smuggling, as well as the actual harm done in particular cases.

The District of Arizona faces especially daunting challenges in combating alien smuggling. The risks inherent in transporting human beings through the harsh and unforgiving desert of Southern Arizona, as well as the increasing violence of the industry,

Statement of United States Attorney Paul K. Charlton - Page 1

have resulted in a disturbing humanitarian crisis. The seriousness of this crisis cannot be underestimated, as illustrated by Arizona District Court Judge James A. Teilborg's emphatic admonition to an alien smuggling defendant that he was "practicing your own form of terrorism." Unfortunately, efforts by federal prosecutors to combat this "trafficking in human misery," as Judge Teilborg calls it, have been hindered by Sentencing Guidelines that inadequately address the level of suffering and risk which defendants involved in alien smuggling impose on their victims. I would like to bring to the subcommittee's attention two recent cases in my district which illustrate some of the inconsistencies and injustices resulting from the lack of sufficiently stringent Sentencing Guidelines for alien smuggling offenses.

In a recent victory, an Assistant United States Attorney successfully prosecuted Aquileo Melchor-Zaragoza, a member of an alien smuggling gang which had been stealing aliens at gunpoint from rival alien smugglers since 1999. After obtaining control over the aliens, the gang members would transport them to a safehouse where the smugglers would threaten both their lives and the lives of their families if their families did not agree to pay for their release. Although no deaths occurred in this case, because the prosecutor was able to charge the defendants with hostage taking and could therefore add counts of brandishing a weapon during and in relation to a crime of violence (18 U.S.C. § 924(c)) to the indictment, Melchor-Zaragoza was sentenced to the significant term of more than 34 years imprisonment. The sentence was a great victory and quite appropriate, not only because of the suffering imposed on the aliens in the smugglers

quest for profit, but because of the increased risk of injury or death the aliens faced as a

result of the smugglers' use of firearms throughout the ordeal.

Unfortunately, in cases where no firearm is used, and thus no hostage taking or

mandatory minimum 924(c) charge is available, it is much more difficult to obtain

significant sentences, although the suffering of the aliens and the risks imposed upon

them by the smugglers' behavior is at least as great as in the case described above, and

often greater in cases where the smuggling results in deaths. The following alien

smuggling tragedy and resulting sentences for those responsible provide a perfect

illustration of this type of injustice.

On May 21, 2001, three men working for a large scale Mexican based smuggling

operation guided a group of twenty-six Mexican nationals across the border into the

United States near Lukeville, Arizona. The guides, including defendant Jesus Lopez-

Ramos, intended to lead the group through the desert to a predetermined point on Arizona

State Highway 85 just north of Ajo, Arizona where they were to be picked up and

subsequently transported to their final destinations at various locations throughout the

United States. Prior to embarking on their journey, the group was told that they would

require only enough water for two days, as their walk through the desert would be

completed in as much time. By the second day, most of the group members had

Statement of United States Attorney Paul K. Charlton - Page 3

exhausted their water supplies. Two of the aliens, along with one of the guides, abandoned the attempt and returned to Mexico. Defendant Lopez-Ramos and another guide, "Lauro," forged ahead, with Lopez-Ramos repeatedly assuring the group that they were a mere "two or three hours away." Lopez-Ramos, however, did not realize that he had taken a wrong turn and was heading not towards the town of Ajo, but away from it, into an uninhabited portion of the desert. Suffering from extreme heat stroke and dehydration, the condition of the aliens rapidly deteriorated. They demanded that the guides leave them to find and return with water. Lopez-Ramos, along with "Lauro," left in an attempt to find water and during that attempt, "Lauro" succumbed to the heat. Lopez-Ramos gave up shortly thereafter and was found by authorities in near-critical condition. On May 23, 2001, Border Patrol agents located five aliens from the group and upon learning that there were others, launched a large-scale rescue operation. In spite of these efforts, however, thirteen of the original twenty-six aliens, as well as the guide, "Lauro," died in the desert as a result of heat-related stress. Eleven other aliens were recovered in critical or near-critical condition suffering from severe dehydration and heat exposure.

Lopez-Ramos pled guilty to twenty-five felony counts of alien smuggling resulting in the deaths of fourteen illegal immigrants and serious bodily injury to eleven others. He was sentenced to 16 years in prison.

Over a period of two years, Francisco Vasquez-Torres, owner and operator of Vasquez Harvesting, a fruit harvesting business located in Lake Placid, Florida, had admittedly been using the Mexico-based smuggling organization responsible for this tragedy to bring Mexican nationals to work for his harvesting business. Vasquez-Torres had conspired and agreed with those in the smuggling organization to bring five of the Mexican nationals from La Paz, Baja, Mexico, to work at Vasquez Harvesting. Vasquez-Torres had, in fact, paid the smuggling fee for each of these five Mexican nationals, with the understanding that the nationals would later repay him from their wages while working for him. Vasquez-Torres's foreman, Joel Viveros-Flores, wired money from Florida to the aliens in Mexico to assist them while they waited to be smuggled into the United States.

Defendant Vasquez-Torres was convicted of conspiracy to bring to the United States for the purpose of commercial advantage or private financial gain certain persons, knowing that said persons were aliens, at a place other than a designated port of entry and at a place other than as designated by the Commissioner of the then-Immigration and Naturalization Service, which resulted in the death of one or more persons, in violation of Title 8, United States Code, Section 1324(a)(1)(A)(i), (a)(1)(A)(v)(I) and (a)(1)(B)(i) and (iv). He was sentenced to 78 months in prison. Defendant Viveros-Flores was convicted of the same charge and sentenced to 18 months in prison.

It is difficult to fathom how in the Melchor-Zaragoza case, where no deaths resulted, a sentence exceeding 34 years was obtained, but in a similar case involving the tragic deaths of fourteen people, and serious bodily injury to eleven others, less than half of that sentence was imposed on Lopez-Ramos, the man who actually carried out the smuggling.  Even more disturbing is the fact that the sixteen-year sentence Lopez-Ramos received was not actually prescribed by the Guidelines, but was instead the result of the prosecutors' request for an upward departure in which they argued that the case fell outside of the "heartland" of cases contemplated by the Sentencing Commission.  The prosecutors requested an eight-level upward departure from the range provided for by the Sentencing Guidelines based on the large number of resultant deaths and physical injuries and the interruption of government functioning caused by the high costs related to the rescue mission.  Indeed, had a departure not been granted in the case, the sentencing range would have been 97-121 months, a maximum of just over 10 years.  As for Vasquez-Torres and Viveros-Flores, the men who provided the financial incentive for the smuggling to occur, mere 78 and 18 month sentences were imposed, respectively. All parties involved were aware of the incredible risks inherent in leading a large group of human beings into a largely uninhabited desert where weather conditions make it nearly impossible to survive a long journey on foot.

I by no means intend to suggest that the sentence in the former case was inappropriate. To the contrary, I believe that sentence to have been quite appropriate.  The

Statement of United States Attorney Paul K. Charlton - Page 6

sentences in the latter case, however, which involved far greater losses to victims, were terribly insufficient.  Presumably, it is the increased risk to the victim which results when a gun is used during the commission of a crime of violence which increases the culpability of the defendant and thus warrants his increased sentence.  The same, if not greater, risks are present each time an alien smuggler brings a group of illegal aliens across the border to embark on an ill-fated attempt to cross the desert, knowing that such attempts result in death in a staggering percentage of cases. Alien smugglers persist in their endeavors in spite of their awareness of this grave risk and do so for the purpose of personal financial gain.  This behavior is at least as culpable as brandishing a gun during the commission of a crime of violence with reckless disregard for the risks to the victims which inhere in such behavior, and I am of the opinion that the Guidelines should reflect this proportionality in culpability.

In addition to the lack of severe penalties for the alien smugglers themselves, the above case illustrates the difficulty with bringing to justice those who do not themselves physically carry out the smuggling, but bear responsibility for its occurrence by seeking the aliens out for labor and paying the smugglers to bring them to the United States in order to work for them.  These individuals are every bit as culpable, if not more, as those guiding the ill-fated journeys across the desert.  Currently, prosecutors must pursue a conspiratorial or aiding and abetting theory of liability in order to obtain any punishment at all for these individuals.  And, as evidenced by the sentences of Francisco Vasquez-

Torres and Joel Viveros-Flores, that punishment is incredibly slight in comparison to the immense amount suffering for which they are responsible.

In addition to this disturbing comparison, countless instances of unjust and inadequate sentences in alien smuggling cases abound. For instance, in the case of *United States v. Miguel and Johnson*, the prosecuting Assistant United States Attorney charged the defendants with three counts of Alien Smuggling Involving Risk of Death, where they had transported three minor children in the trunk of a vehicle during a hot July afternoon in Tucson, Arizona. One of the minors was found unconscious, unresponsive and had to be revived by medical personnel called out to the scene. The defendants pled guilty to the indictment with no plea offer and were sentenced to 21 and 37 months, respectively.

In another case, *United States v. Alderette-Moreno and Loera-Chavez*, two defendants smuggling nineteen illegal aliens instructed the aliens to get into a van, where they were required to lie on the floor of the vehicle and literally pile on top of one another for lack of room, and where the van door was broken and did not close, so that one of the aliens had to hold the door closed while in motion. When one of the van's tires blew, the vehicle rolled, killing the man who had been holding the door, permanently paralyzing another from the neck down, and seriously injuring other occupants of the van. The defendants pleaded guilty, without a plea agreement to Conspiracy to Transport Illegal

Aliens and Alien Smuggling where Death Resulted or Placing in Jeopardy Life of Alien. One defendant was sentenced to 48 months, and the other 57.

Finally, in *U.S. v. Diego Gallegos Castillo,* after walking four to eight hours in the desert to a dry wash area on the Tohono O'Odham Reservation in Sells, Arizona and remaining in the wash for a full day and night, a large group of aliens was picked up by smugglers who instructed them to pile into a pickup truck. Approximately eleven aliens piled into the bed of the truck while three entered the extended cab, which did not have a seat, with a fourteen-year old laying across their laps. While traveling approximately 87-93 mph, the truck flipped over into a wash, resulting in the deaths of four of the aliens. Serious injuries were sustained by the surviving aliens. A jury found the defendant guilty of eleven counts of Transportation of Illegal Aliens Resulting in Death and/or Serious Bodily Injury. The defendant was sentenced to 72 months in prison.

I hope the cases I have presented for your review have been helpful for evaluation of the current state of the Sentencing Guidelines with respect to alien smuggling offenses.

Thank you for your time and attention. I appreciate the opportunity to speak on this important and timely matter. I would be pleased to answer any questions the members might have.

Testimony

of

**Sharon B. Cohn**
**Director of Anti-Trafficking Operations**
**International Justice Mission®**

**Before**
**The Subcommittee on Crime, Corrections and Victims' Rights**
**Committee on the Judiciary**
**United States Senate**

**July 25, 2003**

**Testimony of Sharon B. Cohn**
**Director of Anti-Trafficking Operations, International Justice Mission**
**before**
**The Subcommittee on Crime, Corrections and Victims' Rights**
**Committee on the Judiciary**
**United States Senate**
**July 25, 2003**

Mr. Chairman,

My name is Sharon Cohn and I serve as Director of Anti-Trafficking Operations for International Justice Mission (IJM). On behalf of IJM, I would like to express my thanks to the Committee for the privilege of participating in this important hearing on Alien Smuggling/Human Trafficking: Sending a Meaningful Message of Deterrence. At a time when our nation is vigorously engaged in a struggle against tyranny and terrorism in the world, this Committee manifests the generous and conscientious spirit of the U.S. Senate by making room in its agenda for vigilant oversight of our national commitment to combat the global scourge of human trafficking.

International Justice Mission is an international human rights agency that provides a hands-on, operational field response to cases of human rights abuse referred to us from faith-based ministries serving around the world. Frequently these workers observe severe human rights abuses in the communities where they serve. These workers refer these cases to us, and then we conduct a professional investigation to document the abuses and mobilize intervention on behalf of the victims.

Many of the cases referred to us involve women and children abducted into sex trafficking and commercial sexual exploitation and that is what I will be focusing my testimony on today. IJM deploys criminal investigators to infiltrate the brothels, use surveillance technology to document where the victims are being held, and then identify secure police contacts who will conduct raids with us to release the victims and arrest the perpetrators. We then coordinate the referral of these victims to appropriate aftercare, and support and monitor the prosecutions.

IJM investigators have spent literally thousands of hours infiltrating the sex trafficking industry and working with government authorities around the world to bring effective rescue to the victims and accountability to the perpetrators. In the process, IJM is gaining, I believe, some precise insights about the nature of the problem and helpful lessons about concrete steps that actually prove effective in fighting sex trafficking. We are grateful, therefore, for the opportunity to share something of what we have learned with this Committee.

### Sex Trafficking in the Larger Global Context

While estimates vary, experts agree that between 18,000[1] and 50,000[2] women are trafficked into the United States each year for purposes of commercial sexual exploitation. This level of victimization

---

[1] U.S. Department of State. *Trafficking in Persons Report* (2003).
[2] Website of the U.S. Department of State, Office to Monitor and Combat Trafficking in Persons, viewed July 24, 2003. http://www.state.gov/g/tip/

demands a vigorous response by American law enforcement to intervene, to rescue the victims, ensure their proper care, and to successfully prosecute the perpetrators.

But sex trafficking in the United States cannot be adequately understood in a vacuum. It is imperative in the analysis of trafficking into the United States to consider the global networks which supply the American network. Therefore, we must also turn our attention to the counter-trafficking challenges in the countries from which the victims are trafficked. Cases have been documented of women and children being trafficked into the United States from each of the following countries: [3]

| | | |
|---|---|---|
| Algeria | El Salvador | Mali |
| Armenia | Estonia | Mexico |
| Azerbaijan | Georgia | Nauru |
| Bolivia | Guatemala | Nicaragua |
| Brazil | Haiti | Nigeria |
| Burma | Hungary | North Korea |
| Cambodia | India | Peru |
| Cameroon | Indonesia | Philippines |
| Canada | Italy | Puerto Rico |
| Chile | Jamaica | Romania |
| China | Japan | Russia |
| Colombia | Kiribati | South Korea |
| Costa Rica | Laos | Thailand |
| Czech Republic | Latvia | Ukraine |
| Dominican Republic | Malawi | Uzbekistan |
| Ecuador | Malaysia | Vietnam |

The perpetrator of a crime of human trafficking who is prosecuted in the United States is part of a global chain of illegal transactions that brought the perpetrator and victim into the country. Similar to the drug trafficker caught within the borders of the United States, the sex trafficker is part of a series of transactions which include players in each country of origin, transit, and final destination. In order to effectively disrupt the market, intelligence must be transferred back through this chain of players – from the dealer in the U.S. to his transportation network to his overseas supplier.

In addition, the United States cannot ignore its contribution to the sex trafficking industry. American sex tourists travel worldwide to exploit women and girls. IJM's recent experience in Cambodia demonstrates the multi-national nature of the crime. Vietnamese girls, some as young as 5 years old, are trafficked to Cambodia and exploited by American sex tourists. Of the tens of thousands of women and children in prostitution in Cambodia, the ILO reports that "more than 15 percent of prostitutes were from 9 to 15 years of age, and that 78 percent of these girls were Vietnamese."[4] A survey conducted in December 2001 by World Vision and the Cambodian government indicates that Western pedophiles accounted for about 38 percent of all child sex offenders in Cambodia. Information about the thriving commercial sex trade in Cambodia is easily found by anyone with access to the Internet. Just type the name of a popular Cambodian brothel village like Svay Pak on

---

[3] The Protection Project. A Human Rights Report on Trafficking of Persons, Especially Women and Children (2002).
[4] U.S. Department of State. *Human Rights Report* (2000, 2001 and 2002). Similarly the 2001 Human Rights Report by the U.S. Department of State notes that the International Organization of Migration reported that approximately 3,000 women and girls from southern Vietnam were trafficked into Cambodia, with girls younger than 15 years old constituting 15 percent of this number.

google.com or hop on one of the many pedophile internet chat rooms and you will easily find maps, guidelines, and ratings of the brothels and their services.

More than two years ago, IJM began conducting extensive investigations into one of the most appalling cesspools of child prostitution in the world, a village called Svay Pak outside Phnom Penh where scores of girls between the ages of 5 and 12 were being sold in an open market for pedophiles and sex tourists. Over a two-year period we turned our investigative findings over to Cambodian authorities, but failed to obtain a satisfying response. Then last year, the U.S. Department of State ranked Cambodia as one of the worst offending trafficking countries. The new U.S. Ambassador to Cambodia, Ambassador Charles A. Ray, initiated a very proactive engagement with the senior Cambodian authorities on U.S. policy toward trafficking. This direct advocacy with Cambodian authorities, and the excellent work of Ambassador Ray's staff, helped make it possible for IJM and the Cambodian authorities to bring rescue to 37 minor victims of commercial sexual exploitation out of Svay Pak, including about a dozen children between the ages of 5 and 10. In addition, approximately 16 suspects have been arrested and charged, with cooperative police investigations continuing with IJM to locate and prosecute additional suspects identified in our initial report.

In addition, U.S. Customs is following leads resulting from the raid and we are hopeful that an investigation will result in convictions of identified American sex tourists. I would also like to thank Congress for recently passing the PROTECT Act, paving the way for U.S. Attorneys' Offices to vigorously prosecute Americans who travel abroad and exploit young girls. The Act eliminates the intent requirement and necessitates only that a prosecutor prove an American committed an illicit sexual act abroad. As the little victims in Cambodia told me, many of their clients were Americans. It is imperative that the U.S. crush the demand created by its own citizenry.

## A Market Driven Industry

In a number of countries, IJM has been working hand-in-hand with foreign governments, NGOs and State Department personnel to conduct hands-on operations to rescue victims and to bring perpetrators to justice, and we are learning about the practical impact of effective law enforcement at the street level. Our experience demonstrates that sex trafficking is the ugliest but also the most preventable man-made disaster in our world today.

The simple fact of the matter is this: sex trafficking reaches the United States because it is tolerated by local law enforcement in countries around the world.

This truth is most tragically demonstrated through the lives of the little ones we are privileged to assist in rescuing. A friend of mine, Simla, was raised in a village off a main road to a small city in Asia. When she was 14, a woman in her village sold her to a trafficker. Simla was told she would be working in a noodle shop. When the trafficker brought her to the city, she was sold to a brothel and told that she would have to have sex with customers in order to pay off her debt – a debt that she acquired against her will when the brothel keeper paid the trafficker.

For two and a half years, Simla was subjected to sexual assaults, multiple times a day. She was beaten when she cried, beaten when she was sleepy, beaten when she said she wanted to go home. But the worst beating Simla received, the one that made it difficult for her to walk, was the beating she received the day after a police officer complained that Simla didn't smile after he finished raping her. He would come to the brothel regularly to receive his payment for providing protection for the brothel and sometimes he would rape the girls instead of receiving a cash payment. Simla's friends in the brothel, also children, confirmed that other officers regularly visited the brothel and abused the girls.

IJM investigators identified Simla and others in the brothel as minors and brought our evidence to the attention of the local police. On the night of the raid, a member of the police called the brothel to warn her. The brothel was empty of children by the time it was raided. As the children told us later, they were loaded onto the back of a flat bed truck, covered with a blanket and taken across town. Ultimately, senior police officials communicated down the chain of command that the children must be found and released. Simla and her friends were rescued by the authorities several days later.

Stories like Simla's are repeated throughout the world where local law enforcement do the bidding of the traffickers and brothel keepers. Without police protection, the brothel keeper cannot succeed and with it, he cannot fail. Once the police switch sides, the brothel is fatally vulnerable and effective law enforcement mechanisms can provide rescue and arrests. Until they do, it is the girls that are fatally vulnerable.

This is because sex trafficking is a market-driven industry dictated by predictable economic incentives. It is a straightforward, commercial activity that is predicated upon a very basic risk-to-return calculation. The force of law, when properly applied, can add sufficient risk of criminal sanctions to the traffickers' and brothel keepers' cost calculation, so that the market transaction is no longer worth the threat it represents to their property and liberty.

The business of sex trafficking and commercial sexual exploitation requires that the perpetrators commit multiple felonies of abduction, rape, assault, and false imprisonment – and then it requires that the perpetrators hold out the victims of these crimes openly to the public so that the customers can find them. It does no good at all for the brothel keepers and pimps to hide their victims. In fact, to make money on their investment, the pimps and brothel keepers must make their victims openly available to the customer public – and not just once, but continuously, and over a long period of time. Obviously, therefore, if the customers can find the victims of sex trafficking whenever they want, so can the police. How, therefore, do you possibly get away with running a sex trafficking enterprise? You do so only if permitted by local law enforcement. Generally, this is facilitated by bringing the police into the business and sharing the profits with them in exchange for protection against the enforcement of the laws that are openly and continuously violated every single day the business is in operation. Certainly sex trafficking is exacerbated by poverty and economic desperation; but we do not find epidemic levels of sex trafficking wherever we find poverty in the world. Rather, sex trafficking flourishes on a large scale only in those countries where it is tolerated by local law enforcement.

This is the indispensable insight about the fundamental vulnerability of sex trafficking that must be grasped. Sex trafficking requires the commission of multiple felonies in a way that is held out openly to the public. Therefore it can be shut down wherever there is the political will and operational resources to do so.

### An Intensely Practical Law Enforcement Issue

Accordingly, we commend this committee in raising this issue as one of a critical law enforcement challenge. Tragically, in the international community, the open sale and brutal rape of millions of women and girls is often perceived as a "soft issue." All too frequently, when addressing the issue of sex trafficking overseas, we see the issue communicated through Women's and Children's Bureaus or

Social Services Ministries which though vital, often neglect the deterrent that matters most to traffickers – jail time. Education, awareness, and poverty alleviation programs are important preventative measures, but such programs will never be able to keep pace with the entrepreneurial energy and creativity of the traffickers unless they are combined with practical programs that actually help make national law enforcement successful in sending perpetrators to jail. Police complicity in sex trafficking has been so pervasive and ugly that many have been tempted to imagine solutions that simply ignore the police. But in combating any crime, the answer to *bad* law enforcement is never *no* law enforcement – the answer must always be a committed struggle for *better* law enforcement.

Ultimately the most meaningful message of deterrence is effective law enforcement action. More specifically, the most effective way to disrupt the market for sex trafficking victims is 1) a critical mass of convictions of the perpetrators and 2) accountability for the critical problem of police complicity in countries where trafficking flourishes. From the perspective of the sex traffickers, only two government actions matter: Is the government seriously threatening to actually send me to jail for doing this? And is the government seriously threatening to remove the police protection that I have paid for?

Traffickers, brothel keepers, and pimps are quite willing to endure raids, arrests, and even prosecutions if, at the end of the day, they don't have to actually go to prison. In fact, such actions are just considered part of the cost of doing business. Moreover, even the most corrupt police carry out raids, arrests and initiate prosecutions. In fact, they must do so in order to maintain the credible threat by which they extort bribes from the perpetrators. That is why countries with the worst sex trafficking records can report raids, arrests, and prosecutions; but such countries have very little to report in terms of actual convictions. None of these other actions turn into a credible law enforcement threat that actually deters sex trafficking unless they result in convictions with imprisonment. This is the only cost of doing business that the perpetrators are unwilling to pay.

### The Urgent Need for Cooperative International Relationships

Effective law enforcement also requires expertise and resources. It is critical that the Justice Department share its successes with countries with significant trafficking problems. As with counter-narcotics and counter-terrorism activities, training, resources, collaborative initiatives and anti-corruption efforts are vital to an effective global response to a crime that does not recognize borders.

### Recommendations

Accordingly, I have several specific recommendations I would make to the U.S. government in crafting an effective policy to combat trafficking in the larger global context and send a meaningful message of deterrence to the perpetrators of this abuse.

1. The U.S. government should utilize the federal witness protection program to provide resources to adequately protect cooperating victims' family members who continue to live at risk in source countries.

All law enforcement depends upon the support of the community and the cooperation of the victims. But victims cannot cooperate with law enforcement unless they are provided with a safe environment in which they feel empowered to participate in the justice system. In contrast, treating victims as criminal disables law enforcement efforts. The United States has an opportunity to set a standard for the world by the way we treat women and girls who are trafficked into this country. By employing the T visa and utilizing witness protection mechanisms, authorities can create a safe, non-coercive environment for the victims – an environment that vastly enhances the chances of their cooperating in the prosecution of the criminals. Protection for these victims' families abroad is also of primary concern to victims. International cooperation is essential to extend the witness protection program to those family members abroad who are still vulnerable to violent reaction by the hands of the criminal network. Equally critical are programs that fund comprehensive and secure aftercare services for the victims of sex trafficking. Victim protection is a need that can and must be addressed by targeted and generous appropriations.

2. The Attorney General should vigorously utilize the PROTECT Act to prosecute American sex tourists.

Congress recently passed the PROTECT Act which frees the U.S. Attorney Offices to vigorously prosecute sex crimes committed by Americans abroad. We must encourage prosecution and conviction of these criminals and subsequent media coverage to send a deterrent message to would-be criminals. For the sex tourist and pedophile, the cost of the risk of arrest and conviction must become prohibitively high.

3. Federal law enforcement agencies and the Department of Justice should continue to communicate through joint training initiatives and funding that sex trafficking is a priority issue and a violent crime worthy of the attention of elite law enforcement.

Sex trafficking can be drastically reduced by effective, concerted law enforcement action. The quality and vigor of local law enforcement's response to commercial sexual exploitation is driven by the priorities of senior level political authorities; the quality of resources and training provided to local law enforcement; and the clarity and comprehensiveness of the law. Sex trafficking is not a soft issue, nor a crime of mere vice, but a violent crime that requires multiple felony offenses. It deserves treatment as a priority issue worthy of the attention of elite law enforcement forces.

However, even urgent law enforcement priorities cannot be vigorously and effectively pursued without resources and training initiatives that equip street level enforcement to be effective. Governments need the practical wherewithal and the operational capacity to take decisive law enforcement measures to combat trafficking and to care for the victims. Accordingly, funding is necessary to strengthen law enforcement capacities to investigate, arrest, and prosecute sex trafficking offenders. Programs are needed to support special anti-trafficking police units and prosecutorial teams with training, operational support, and hands-on assistance in achieving the priority outcome of sending offenders to jail and removing dirty cops. The FBI Academy in Bangkok provides a course on "Illegal Migration and Trafficking in Women and Children" as part of its core training. Courses like this should be expanded and offered in other key trafficking locations.

In addition, the U.S. should demonstrate its commitment to combat trafficking by providing resources to local law enforcement in high-trafficking areas. Funding for salaries and supplies in conjunction with training assistance will further U.S. counter-trafficking goals.

4. Because the very nature of sex trafficking cases makes them multi-national, the U.S. government should continue to improve information sharing and collaboration both among U.S. agencies and foreign governments.

Successful prosecutions require information sharing. Federal law enforcement officials should obtain intelligence from trafficking victims in this country about how they were trafficked, from what city and by what means. This intelligence should be regularly forwarded to law enforcement attaches at U.S. embassies and made available to local and national law enforcement authorities of foreign source countries. This will facilitate the interdiction of traffickers from the source country and disrupt the market.

It is our belief and expectation that the majority of sex trafficking can be eliminated where the police have switched sides and serve the public rather than the traffickers and brothel keepers. The U.S. government should continue to provide encouragement, expertise, and resources to that end.

Thank you again, Senator Graham, for calling attention to this important matter. It is an honor to speak before the Committee today.

---

STATEMENT

OF

# CHARLES H. DEMORE
### INTERIM ASSISTANT DIRECTOR OF INVESTIGATIONS

FOR

**BUREAU OF IMMIGRATION AND CUSTOMS ENFORCEMENT
THE DEPARTMENT OF HOMELAND SECURITY**

**REGARDING A HEARING ON**

## ALIEN SMUGGLING

**BEFORE THE**

**SENATE SUBCOMMITTEE ON CRIME,
CORRECTIONS AND VICTIMS' RIGHTS**

**COMMITTEE ON THE JUDICIARY**

**JULY 25, 2003
10:00am
ROOM 226 SENATE OFFICE BUILDING**

MR. CHAIRMAN AND MEMBERS OF THE COMMITTEE, thank you for the opportunity today to address you regarding the efforts of the Bureau of Immigration and Customs Enforcement (BICE) to combat the smuggling of aliens into the United States.  I am Charles DeMore, the Interim Assistant Director of Investigations and I am pleased to have the opportunity to share my experience and knowledge with you regarding this important issue.

The creation of the new Department of Homeland Security (DHS), and specifically BICE, combined legal authorities and investigative tools to effectively combat organized human smuggling and trafficking by investigating and administratively prosecuting immigration violations related to criminal organizations involved in smuggling, transporting, and harboring of aliens; money laundering; racketeering violations; human trafficking and child forced labor provisions.  In addition, the new BICE structure provides a more effective means of dismantling and disrupting the criminal activities of these organizations, with tools such as financial and data analysis, telecommunication intercepts, and air and marine interdiction capabilities.  However, no enforcement effort would be complete without the cooperation and collaboration of the Bureau of Customs and Border Protection (BCBP).  I am pleased to tell you that we are fully engaged with our partners in both BCBP and other DHS components in combating these crimes.

I would like to begin by providing an important clarification and necessary distinction between the terms alien smuggling and human trafficking.  Alien smuggling and human trafficking, while sharing certain elements and attributes and overlapping in some cases, are distinctively different offenses.  Human trafficking, specifically what U.S. law defines as "severe forms of trafficking in persons," involves (unless the victims are minors trafficked into sexual exploitation) force, fraud or coercion, and occurs for the purpose of forced labor or commercial sexual exploitation.  Alien smuggling is an enterprise that produces short-term profits resulting from one-time fees paid by or on the behalf of migrants smuggled.  Trafficking enterprises rely on forced labor or commercial sexual exploitation of the victim to produce profits over the long-term and the short-term.

Smugglees are willing to risk potential death seeking their dream and are normally free to seek it once they reach their final destination.  On the other hand, we know that trafficking victims find themselves in a servitude arrangement that does not end once they have reached their destination, and further may find themselves moved from one destination to another against their will.

Human smuggling has become a lucrative international criminal enterprise and continues to grow in the United States.  This trade generates an enormous amount of money - globally, an estimated $9.5 billion per year.  The commodities involved in this illicit trade are men, women, and children. Traffickers or smugglers transport undocumented migrants into the U.S. for work in licit, semi-illicit and illicit industries.  The traffickers' foremost goal, like the smuggler, is to maximize profits.  The sale and distribution of smuggled humans in the U.S. is a global, regional, and national phenomenon.  Women and children are trafficked short distances within the U.S.

(small towns to bigger cities), as well as coming from as far away as China, Ukraine and Thailand.

The U.S. Department of State has estimated that at any given time, there are hundreds of thousands of people in the smuggling pipeline, being warehoused by smugglers, waiting for new routes to open up or documents to become available -- and their primary target is the United States.

While human trafficking cases have attracted media attention, the loss of life in an alien smuggling case is no less tragic.  To illustrate the callous disregard smugglers have for human life I would like to provide you with the details of some tragic incidents involving deaths (noting that some of these smuggling cases may be trafficking cases as well):

**Iowa** – In October 2002, 11 undocumented aliens were found dead in a covered grain car near Dennison, IA.  It was determined that they had been smuggled and their bodies trapped in the grain car for four months.  This crime is the subject of an ongoing investigation.

**Texas** – In May 2003, 17 undocumented aliens were found dead inside a tractor-trailer in Victoria, Texas.  Four hours into their 300-mile trip to Houston, oxygen ran out in their dark, sealed, hot, airless trailer.  These aliens had beat their way through the trailer taillights in a desperate attempt to signal for help.  Within 72 hours of the discovery, the collective efforts of Special Agents and intelligence analysts from BICE, our counterparts in the Bureau of Customs and Border Protection, the Texas Department of Public Safety, the Victoria County District Attorney's Office, the United States Secret Service, and the Victoria County Sheriff's Office, led to the identification and arrest of four defendants in Ohio and Texas.  As of July 17, 2003, a total of fourteen defendants have been charged with various smuggling-related crimes arising from this tragic incident.  The successes that we achieved in this operation are a direct result of fully integrating BICE special agents and other personnel, equipment and methodologies into a unified law enforcement effort.  Still, the smugglers remain undaunted by the tragedy.  They continue to use sealed railroad cars and tractor-trailers to move illegal aliens through the South Texas smuggling corridor.  In fact, only days after the grisly discovery in Victoria, Texas, 16 other migrants were discovered in a tractor trailer only an hour away.

**Washington** – In January 2000, three undocumented aliens were found dead in the cargo container of a vessel in Seattle, WA.  The three were part of a group of eighteen smuggled Chinese aliens that had been sealed in the container for a period of two weeks.  The survivors, who were in dire medical condition, remained in the container with the deceased until their discovery.

**California** – In March 2000, six undocumented aliens were found in the San Diego east county mountains, four of which died due to hypothermia.  The smugglers abandoned the group in the snowy mountains as the aliens pleaded not to be stranded.

**Florida** – In December 2001, a capsized vessel was found in the Florida Straits, alleged to have been carrying 41 Cuban nationals, including women and children.  All are believed to have perished at sea.

**New York** – In June 1993, the Golden Venture, a vessel that had traveled 17,000 miles in 112 days from China, ran aground off the coast of Queens in New York City. The human cargo suffered subhuman living conditions during the voyage with inadequate food and ventilation. Most of the 286 people jumped into the frigid Atlantic Ocean, 10 of whom drowned.

**Arizona** – In 2002, 133 deaths were recorded relating to alien smuggling loads in the Arizona deserts. Tragically, many of these deaths were due to aliens being abandoned and lost in the desert heat, but some of these deaths were homicides. The BICE Phoenix Special-Agent-in-Charge is currently involved in an investigation in which as many as 13 homicides have been attributed to alien smuggling. Several of the deceased were undocumented aliens who were unable to pay their smuggling fees. Local law enforcement agencies attribute most of the increase of violent crime, hostage taking, and home invasions in Arizona as being related to alien smuggling.

As you can see, alien smuggling is not confined to any geographic region; it is a problem of national scope, which requires a coordinated national response. BICE is developing a foreign and domestic strategy, which includes the implementation of critical incident response teams. The purpose of these investigative teams is simple and effective: begin the investigation of a critical incident as quickly as possible, assembling the broad spectrum of technical and subject matter expertise that is needed to solve complex investigations.

The teams will consist of Special Agents drawn from BICE assets who possess specialized skills in the full constellation of investigative techniques; language and cultural skills, land, air and maritime smuggling, crime scene management, technical operations and forensics. BICE Victim-Witness Coordinators would supplement the teams as needed. This investigative response will be coordinated at a proposed BICE Smuggling Coordination Center utilizing resources and equipment deployed in key geographic areas nationwide.

Increased efforts are also being placed on addressing the smuggling of juveniles into the United States, which has surged in recent years. This increase is driven by the demand created by U.S. citizens wanting to illegally adopt children from abroad, immigrants attempting to reunite their families, and child exploitation. Mexican consulates in Southern Arizona alone handled more than 1,500 repatriations of unaccompanied Mexican juveniles during the first half of 2002.

In contrast to the smuggling of family members, trafficked children are often lured by promises of education, a new skill or a good job; other children are kidnapped outright, taken from their home villages or towns and then bought and sold as commodities. Attracted by enormous profits and minimal risks, criminal organizations at all levels of sophistication are involved in the trafficking of children as human cargo across international borders for sexual exploitation and forced labor.

The fall of communism, coupled with the deteriorating third world economies, has fueled the dramatic rise of this sinister form of commerce. Additionally, international organized crime groups such as the Chinese Triads; Japanese Yakuza; Russian, Albanian, Georgian, Ukrainian, Polish, Nigerian, and Thai criminal networks have also capitalized on weak economies; corruption, and improved international transportation infrastructure in order to facilitate the smuggling and trafficking of some 700,000 to 2,000,000 people globally each year. Some of these organizations have abandoned their historic ethnic alliances to join together in criminal enterprises and to hinder U.S. Government law enforcement efforts.

The national and international enforcement environment changed significantly after the September 11 attacks. BICE places a significant emphasis on targeting alien smuggling organizations that present threats to national security. This emphasis recognizes that terrorists and their associates are likely to align themselves with specific alien smuggling networks to obtain undetected entry into the United States. In addition to the emerging terrorist threat, three factors have created an environment in which terrorists and smuggling enterprises may combine their criminal efforts to pose a significant national and international threat. These factors are:
1) The involved criminal organizations growing volume and sophistication,
2) Their ability to exploit public corruption; and,
3) Lax immigration controls in source and transit countries.

As in our war on terrorism, the most effective means of addressing these issues is by attacking the problem in source and transit countries thereby preventing entry into the United States. For many years, we have recognized the need to identify and dismantle large-scale trans-national smuggling organizations and have done so in collaboration with other law enforcement agencies, both foreign and domestic. And we have served as co-chair, with the Department of Justice and the Central Intelligence Agency, to an interagency working group on smuggling targeting criminal organizations that present national security concerns for the United States. In the aftermath of September 11 and in concert with the intelligence community, we redirected our efforts to focus on smuggling organizations alleged to smuggle aliens who have ties with terrorists groups. Consequently, BICE is developing a strategy that will address alien smuggling and human trafficking at the national and international levels. The global Anti-Smuggling/Human Trafficking Strategy will concentrate our efforts in intelligence-driven investigations against major violators, specifically targeting organizations with ties to countries that support terrorist organizations such as al Queda.

Members of this Subcommittee have previously raised the issue of the need for enhancing the penalties for smuggling offenses. While we believe the penalties set forth in section 274 of the Immigration and Nationality Act to be adequate, in practice the sentences imposed in cases have traditionally been quite short. Sentencing enhancements mandated by the Illegal Immigration Reform and Immigrant Responsibility Act of 1996 have, however, resulted in significant increases in sentences imposed in the past few years.

We look forward to working with this Committee in our efforts to save lives and secure our national interests.  I hope my remarks today have been informative and helpful to each of you in understanding the complexity surrounding these issues. I thank you for inviting me to testify and I will be glad to answer any questions you may have at this time.

Statement of


# ROBERT L. HARRIS

**DEPUTY CHIEF, UNITED STATES BORDER PATROL
BUREAU OF CUSTOMS AND BORDER PROTECTION
THE DEPARTMENT OF HOMELAND SECURITY**


Before the

**UNITED STATES SENATE
COMMITTEE ON THE JUDICIARY
SUBCOMMITTEE ON CRIME, CORRECTIONS AND VICTIM'S RIGHTS**


**Regarding
"Alien Smuggling/Human Trafficking: Sending a Meaningful Message of
Deterrence"**

**July 25, 2003
10:00 a.m.
Room SD-226, Dirksen Senate Office Building**

Chairman Graham, Ranking Member Biden, and distinguished Subcommittee Members, it is my honor to have the opportunity to appear before you today to discuss efforts to prevent and deter the illegal entry and smuggling of undocumented aliens into the United States through the operations and law enforcement initiatives of the United States Border Patrol, now a component of the newly created Bureau of Customs and Border Protection.

My name is Robert L. Harris, and I am the Deputy Chief of the United States Border Patrol. I would like to begin by giving you a brief overview of our agency and mission.

As you know, on March 1, 2003, Immigration Inspectors and the U. S. Border Patrol from the Immigration and Naturalization Service (INS), Agricultural Inspectors from the Animal and Plant Health Inspection Service (APHIS), and Customs Inspectors from the U. S. Customs Service merged to form the Bureau of Customs and Border Protection (BCBP) within the Border and Transportation Security (BTS) Directorate of the Department of Homeland Security. Now, for the first time in our country's history, all agencies of the United States government with significant border responsibilities have been brought together under one roof. With our combined skills and resources, we will be far more effective than we were when we were separate agencies.

Within BCBP, the mission of the Border Patrol remains virtually unchanged. We are responsible for providing Homeland Security along our Nation's borders between ports of entry. Through our operational Sectors, we patrol and secure 4,000 miles of international land border with Canada and 2,000 miles of international land border with Mexico. We also patrol roughly 2,000 miles of coastal waters surrounding the Florida Peninsula and Puerto Rico. We do this with over 10,400 Border Patrol Agents. nationwide. While our priority mission is to detect and prevent terrorists and terrorist weapons, including weapons of mass destruction. from entering the United States between the ports of entry, we also interdict illegal migrants and contraband.

Illegal migration and alien smuggling into our country is a serious problem to those who live and work in the border community. but its impact and the associated criminal activity that accompanies it is far-reaching. An uncontrolled border presents great concern, spreading border violence, and degrading the quality of life in border communities and other locations affected by this activity.

Since 1994, the U.S. Border Patrol has operated under a comprehensive national strategy designed to gain and maintain control of our Nation's borders. Major initiatives such as *Operation Hold the Line*, in our El Paso Sector, *Operation Gatekeeper* in our San Diego Sector, and *Operation Rio Grande* in our McAllen Sector have had a significant effect on illegal migration along the Southwest Border. These initiatives sought to bring the proper balance of personnel, equipment, technology and infrastructure into areas experiencing the greatest level of illegal activity on the southwest border. In the ensuing years, our operational manpower has more than doubled. Enforcement related technology has been applied to support our agents, especially in isolated and remote areas of the border.  Existing resources such as air and marine units, horse patrols and all terrain vehicles have been enhanced to support day-to-day field operations.  Infrastructure has been deployed in the way of fencing, vehicle barriers, cameras and lighting to assist field agents in their efforts to deter and prevent the flow of illegal aliens and contraband. While there is certainly more that must be done in this area, the strategy has yielded important results.

In the wake of 9-11, vulnerabilities and deficiencies along the northern border have received increased attention, challenging us to increase our enforcement presence along the northern border.  With the recent reassignment of more than 375 agents to the northern border, there will be 1,000 agents strategically and permanently placed along the northern border by the end of the year, enforcing a northern border strategy built on interagency and international cooperation and coordination, effective technology development and deployment, and innovative resource allocation.

Overall, our efforts have been very successful, with decreases in apprehensions and illegal entries, indicating that our efforts have had an impressive deterrent effect. Alien apprehensions have declined from a high of 1.6 million in fiscal year 2000, down to a 28-year low of less than 1 million in fiscal year 2002.  In fiscal year 2001, the Border Patrol arrested 1.26 million aliens on all borders, which was a 24% decline from the previous year.  In fiscal year 2002, 955,310 aliens were arrested, which represented a 25% decrease from fiscal year 2001.  Border Patrol is also among the leaders on the Southwest border in narcotics seizures.  In fiscal year 2002, we seized 1,211,009 pounds of marijuana, and 14,463 pounds of cocaine.  Through it all, the Border Patrol has maintained and encouraged a positive relationship with local communities, including ranchers, farmers and other law enforcement entities.  The crime rate along the Southwest Border, where we have had significant operations, paralleled the decline in apprehensions.  Places like San Diego, El Paso, and McAllen have experienced decreased crime rates, increasing safety for our agents and the local population, and improving the overall quality of life in those areas.

Over the past several years, unscrupulous alien smugglers have moved migrants into more remote areas with hazardous terrain and extreme conditions.  As smuggling tactics and patterns have shifted, our strategy has been flexible enough to meet the challenges head on.  Building on longstanding public safety and humanitarian measures practiced by the U.S. Border Patrol, we have implemented initiatives to increase border safety along the Southwest Border and we have taken steps to enhance our levels of preparedness.  We have developed a Border Safety Initiative (BSI) with the goals of reducing injuries and preventing deaths along the southwest border region.  Striving to create a safer border environment, BSI is not only proactive in informing potential migrants of the hazards of crossing the border illegally, but also provides quick response to those who are in life-threatening situations.  Working with local television, radio and newspaper agencies, both in the United States and Mexico, we have developed and delivered public service announcements and advertisement campaigns to increase public safety awareness, and to educate the public regarding our mission, benefiting our law enforcement efforts.

Border Patrol maintains a number of 24-hour checkpoint operations in many of our Sectors, and they are an integral part of border control. Their strategic placement and operations serve to prevent and disrupt alien and narcotic smuggling, and provide increased control and deterrence at the border. The presence of a checkpoint forces smugglers and illegal entrants to change their entry and travel patterns -- to our tactical advantage. A sustained border enforcement presence, supported by checkpoints that screen traffic traveling away from the border, adds an additional level of security nationally. Checkpoints are an essential part of border enforcement--the significant number of drug seizures and alien apprehensions at checkpoints clearly demonstrate this. What cannot be measured is the significant deterrent effect these operations have on smuggling and other illegal activity.

The challenges we face with existing infrastructure at our checkpoints will continue to be addressed in an effort to update, expand and modernize, and we will continue to work diligently under the limitations that now exist. With an ever-increasing volume of traffic, agents have mere seconds to conduct immigration checks, and to decide if probable cause exits to warrant additional inspection. Technology will continue to play a key role in this effort. After the recent and notoriously tragic smuggling case in Victoria, Texas, the Bureau of Customs and Border Protection Commissioner Bonner expedited the deployment -- within 72 hours of the incident -- of mobile truck x-ray machines to South Texas Border Patrol checkpoints. This technology, long in use at ports of entry, will be an added asset to the Patrol in combating the work of smugglers. The deployment of this technology would not have been as easily undertaken without the creation of DHS and the merging of border enforcement agencies within the Bureau of Customs and Border Protection.

While the Border Patrol provides a significant law enforcement presence on the border, we are also recognized as a major source of information and intelligence, including on matters relating to National security risks.  For example. our Intelligence Section provides valuable information regarding apprehended aliens. and since the reorganization, it has been fused with the Intelligence assets of the BCBP.  Recognizing that border security cannot be a singular effort, but a collaborative, multi-agency effort; we coordinate our efforts. disseminate information, and share intelligence with other federal, state and local law enforcement agencies, strengthening the cord of better enforcement, better intelligence and better security.

Nationally, the Border Patrol is tasked with a very complex, sensitive and difficult job, which historically has presented immense challenges.  Homeland security has become a top priority.  The Border Patrol is proud to be the "front line" of defense for this very important mission.  The challenge is huge, but one which we face everyday with vigilance, dedication to service, and integrity.

I would like to thank the Subcommittee for the opportunity to present this testimony today, and I would be pleased to respond to any questions that you may have at this time.

### #

# News Release

# JUDICIARY COMMITTEE

*United States Senate • Senator Orrin Hatch, Chairman*

uly 25, 2003                                   Contact:  Margarita Tapia, 202/224-5225

**Statement of Senator Orrin G. Hatch**
**Before the United States Senate Committee on the Judiciary Subcommittee on Crime,**
**Corrections and Victims' Rights**
**Hearing on**

## "ALIEN SMUGGLING/HUMAN TRAFFICKING: SENDING A MEANINGFUL MESSAGE OF DETERRENCE"

I want to thank Chairman Graham for holding this very important hearing before the Crime, Corrections and Victim's Rights Sub Committee.  Alien smuggling and human rafficking are crimes that carry with them the most devastating of consequences for victims. ust this year we were all shaken by the terrible tragedy in South Texas where heartless and lepraved smugglers abandoned men, women and children in the back of a tractor-trailer rig in earing heat with no air.  Press reports indicate that the victims clawed at the trailer in an attem o get air, escape and survive.  The victims ranged from a 7-year-old boy to a 91-year-old man.

*The Washington Post* reports that, over the past five years, more than 2,000 other nigrants have died attempting to enter the United States, often of exposure in the hot Arizona leserts, freezing to death in mountains, or drowning in the Rio Grande.  Smugglers take idvantage of individuals from other countries whose situation is so desperate, that they feel lik hey have no other choice but to sell all of their belongings and subject themselves to a smuggl vho promises them transportation to this country.

This type of crime is no longer limited only to Border states.  No state is immune from he affects of smuggling and trafficking.  As stepped up enforcement occurs along the tradition muggling routes, smugglers move their cargo, human beings, through other routes into the nterior of this country.  In my own state of Utah, Salt Lake City is at the crossroads of Intersta .5 which extends up from the south, and Interstate 80, which extends east and west.  This iighway crossroads has become an area of opportunity for alien smuggling as aliens move fron iur Border States into our interior States.  Smugglers often use vehicles that are unsafe, :ndangering the lives of the aliens -- and the general public as well.

The recent examples of smuggling deaths highlight the callous disregard smugglers and raffickers have for what they view solely as cargo or chattel.  These examples illustrate the ne o make sure that our criminal justice system does all it can do to deter those who smuggle and raffic in human beings by punishing them to the full extent authorized by law.

I look forward to hearing from our witnesses today about this important issue.

# # #

# U S  SENATOR PATRICK LEAHY

CONTACT: David Carle, 202-224-3693                    **VERMONT**

**Statement of Senator Patrick Leahy**
**Ranking Member, Senate Judiciary Committee**
**Subcommittee on Crime, Corrections, and Victims' Rights**
**Hearing on "Alien Smuggling/Human Trafficking:**
**Sending a Meaningful Message of Deterrence"**
**July 25, 2003**

This hearing addresses important topics that have become an increased focus of public attention in recent years – alien smuggling and human trafficking.

I am proud to have served on the conference committee that produced the Victims of Trafficking and Violence Protection Act ("VTVPA") of 2000.  On the Senate side, this bill was the handiwork of Senators Brownback and Wellstone, and it typifies Senator Wellstone's commitment to protecting the downtrodden.  In addition to providing services for trafficking victims and providing for a system of incentives and sanctions designed to reduce trafficking worldwide, the law increased criminal penalties for existing offenses and created new offenses, including forced labor, trafficking with respect to peonage, slavery, involuntary servitude, or forced labor, and sex trafficking of children.  I look forward to hearing whether the law is aiding prosecutors, and whether further legislation is needed in this area.

The VTVPA created the T and U visas to aid immigrants who were victimized.  The T visa is available to severe victims of trafficking.  The U visa is available to immigrants who were the victims of certain listed crimes, most of which are gender-related.  To receive a U visa, aliens must show that they have been helpful, are being helpful, or are likely to be helpful to law enforcement.  Unfortunately, the processing of the U visas has been disorderly at best.  First, although the visa took effect upon enactment of the law, implementing regulations have still not been issued.  Second, adjudications of U visa petitions have not been centralized, unlike petitions for T visas or immigration relief under the Violence Against Women Act, which are processed at the Violence Against Women unit ("VAWA unit") at the Vermont Service Center.  As a result, the adjudication of U visa petitions has been inconsistent and seemingly deserving applicants have been denied relief.  I am very proud of the work the VAWA unit has done with gender-related immigration petitions, and I filed an amendment to the DHS appropriations bill that would have given the unit responsibility for adjudicating U visas as well.  I am pleased that after I filed the amendment, the Department of Homeland Security ("DHS") informed my office that the adjudication of U visa petitions will be centralized at the VAWA unit.  This would be a great victory for immigrant victims of crime.

senator_leahy@leahy.senate.gov
http://leahy.senate.gov/

Alien smugglers, like human traffickers, exploit the intense desire of many around the world to come to the United States. We have seen in recent years a series of tragic deaths of aliens who were piled into airless compartments. Most recently, 19 aliens – from Mexico, El Salvador, Guatemala, Nicaragua and Honduras – died in May in Victoria, Texas. The Washington Post has reported that more than 2,000 aliens have died in the last five years in attempts to cross our border. A substantial portion of them undoubtedly had cast their lot with smugglers. I am pleased that significant law enforcement resources were devoted to the Victoria deaths, with an indictment returned the next month. I hope also to hear what resources are being devoted to less-publicized smuggling cases, and of ongoing efforts to reduce the incidence of smuggling in the first place.

# Department of Justice

STATEMENT

OF

JOHN MALCOLM
DEPUTY ASSISTANT ATTORNEY GENERAL
CRIMINAL DIVISION
DEPARTMENT OF JUSTICE

BEFORE THE

SUBCOMMITTEE ON CRIME, CORRECTIONS, AND VICTIMS RIGHTS
COMMITTEE ON THE JUDICIARY
UNITED STATES SENATE

CONCERNING

ALIEN SMUGGLING AND HUMAN TRAFFICKING:

TWO DISTINCT CRIMES POSING CHALLENGES
FOR INTERNATIONAL LAW ENFORCEMENT

PRESENTED ON

JULY 25, 2003

STATEMENT OF JOHN MALCOLM
DEPUTY ASSISTANT ATTORNEY GENERAL
UNITED STATES DEPARTMENT OF JUSTICE
CRIMINAL DIVISION

BEFORE THE SENATE SUBCOMMITTEE ON
CRIME, CORRECTIONS AND VICTIMS' RIGHTS
COMMITTEE ON THE JUDICIARY
UNITED STATE SENATE

ALIEN SMUGGLING AND HUMAN TRAFFICKING:

TWO DISTINCT CRIMES POSING CHALLENGES FOR
INTERNATIONAL LAW ENFORCEMENT

July 25, 2003

Mr. Chairman, Senator Biden, it is my privilege to appear today before this Subcommittee

to discuss the problems of international alien smuggling and trafficking in persons, also referred

to as human trafficking.  These two serious crimes – distinct in their nature but related in their

effects – are of great importance to the Department of Justice, because they present both national

security and human rights concerns.  Ultimately, alien smuggling and human trafficking subvert

our sovereignty.  Alien smuggling puts the decision about who enters our country into the hands

of criminals, who may not know and probably do not care if their actions help terrorists or other

criminals to enter our country.  Both alien smuggling and trafficking in persons strain limited

resources and penalize persons who wish to enter our country legally.  These crimes also enable

international criminal organizations to flourish throughout the world, and breed corruption --

often of border officials in other countries – thereby undermining respect for the rule of law and

harming basic democratic institutions.

Some have argued that alien smuggling is a so-called "victimless" crime.  I would like to

put that pernicious myth to rest.  Smuggled migrants are often subjected to violence and inhumane and dangerous conditions. Some are trafficked into sexual exploitation or forced labor. Others die every year from drowning, abandonment, accidents or brutality by smugglers. While it is not uncommon to find one or two bodies in the mountains or in the desert, it is only when a large number of people die that the danger that smuggled migrants face garners national attention. Even if migrants arrive at their intended destinations alive and unscathed, however, smugglers have been known to extort the payment of their exorbitant fees by forcing migrants into virtual slavery, including selling them into sexual exploitation, or by holding family members back in the home country for ransom.

<u>Legislative and Law Enforcement Efforts Over The Past Several Years</u>

By the early 1990s, the Department of Justice recognized that alien smuggling was becoming a major international organized criminal activity, and that additional law enforcement tools were needed to address the problem. With Congress's assistance, we were able to obtain additional statutory tools which significantly enhance our ability to enforce criminal immigration laws. Over the last ten years, Congress has increased the statutory maximum penalties for alien smuggling and immigration document fraud, and the United States Sentencing Commission has increased the applicable penalties under the federal sentencing guidelines. Alien smuggling-related offenses were made predicate offenses for RICO and money laundering, and were also added to the list of offenses for which investigators can obtain court-authorized interception of wire, oral, or electronic communications. Additionally, the assets of alien smugglers are now subject to civil and criminal forfeiture.

Using these tools, the Department of Justice prosecutes a large number of alien smuggling cases every year. In Fiscal Year 2001, approximately 1,900 defendants were convicted of alien smuggling offenses, 17 percent of the total number of immigration-related convictions that year.

More recently, Congress greatly improved the Department's ability to prosecute human trafficking with the passage of the Trafficking Victims Protection Act (TVPA) of 2000. In enacting this ground-breaking new law, Congress recognized the growth and devastating impact of trafficking in persons around the globe. It saw that trafficking flourished where political turmoil and economic dislocation made people desperate for opportunity and, consequently, susceptible to the threats and schemes of traffickers promising passage to a better life. Congress also noted that trafficking flourishes because it is a profitable and relatively low-risk enterprise. Relying on the TVPA and a variety of other statutes, the Civil Rights and Criminal Divisions have moved forcefully to punish traffickers and to assist victims. As of July 15, 2003, there were approximately 122 open trafficking investigations – nearly twice as many as were open in January 2001. Over half of these have been initiated as a result of the Department of Justice's Trafficking in Persons and Worker Exploitation Task Force Complaint Line, established in February 2000, and many involve charges under the TVPA. This Act added new crimes, outlawing forced labor, sex trafficking of children, or of adults by force, fraud, or coercion, and document abuse. Together with pre-existing trafficking statutes such as 18 U.S.C. § 1584, which enforces the 13th Amendment's proscription against involuntary servitude, the TVPA enables the Department to prosecute those crimes in which threats of serious harm, physical restraint, or abuse of the legal process serve to obtain or maintain the labor or services of a trafficking victim, or where traffickers use force, fraud or coercion to cause their victims to engage in commercial sex. The Act also doubled the possible sentences that traffickers can receive. Prosecutorial guidance was sent to the field regarding these new crimes, and the U.S. Sentencing Commission

has increased the applicable penalties accordingly.

<u>Nature Of The Offenses</u>

Smugglers have been assisting people to cross illegally into the United States for as long

as immigration has been regulated.  Smugglers often conspire with traffickers essentially to

enslave migrants once they arrive in the United States.  However, it is only within the past fifteen

years that we have seen the rise of organized, international alien smuggling and trafficking in

persons networks.  In addition to posing national security threats, some of these networks are

highly sophisticated and generate millions of dollars.  Globally, according to some estimates,

alien smuggling and trafficking in persons generate billions of dollars per year.  The demand

created by people who seek illegal entry to work in the United States (and a handful of other

desired destination countries) has created a lucrative market for smugglers, traffickers, and

document vendors to exploit.  In addition to being highly profitable, alien smuggling and human

trafficking are relatively low risk when compared to other illicit conduct.  One reason is that the

laws and enforcement capabilities of many source and transit countries are weak or non-existent

with regard to alien smuggling and document fraud.  Often, the principals in smuggling

organizations and networks never physically enter the United States (or other destination

countries) themselves.  In contrast, while traffickers often enter and reside in the United States,

some disappear into ethnic communities and, once so ensconced, turn to further exploiting

members of those ethnic communities.  Whether operating within our borders or abroad, alien

smugglers and human traffickers both operate in the netherworld of illegal aliens.

<u>Evolution Of Trafficking And Smuggling Networks</u>

In the early 1990's, prior to Congress's increasing penalties for alien smuggling, we noticed that some of the Asian street gangs in New York City, San Francisco, Boston and elsewhere were rapidly expanding their alien smuggling operations, precisely because of the high returns and perceived low risks. A string of successful RICO prosecutions, which focused in part on violent activities linked to alien smuggling -- such as extortion and hostage taking -- broke the backs of the Chinatown street gangs. Unfortunately, it appears that the void has been filled by newer, more sophisticated smuggling networks.

Human trafficking is evolving in the wake of these sophisticated alien smuggling networks, but the Department is responding in kind. Using the new tools created by the TVPA, the Department has been able to reach a variety of trafficking schemes, falling into four main categories: trafficking for purposes of sex; forcing individuals to perform domestic service; forced labor in factories; and forced agricultural labor. Such cases have ranged from large-scale, multiple victim prosecutions, such as *United States v. Kil Soo Lee*, where over 200 victims were enslaved in a garment factory in American Samoa, to individual victims in cases which nonetheless have international implications, such as *United States v. Blackwell*, where a domestic servant was forced to perform labor through threats by, among others, a high-level Ghanaian government official.

<u>Investigating And Prosecuting Alien Smuggling And Human Trafficking</u>

The investigation and prosecution of both alien smuggling and trafficking cases can come about in a number of different ways. Some investigations are triggered by the discovery of smuggling operations or trafficking in persons situations in progress, such as the ongoing alien smuggling investigations and prosecutions stemming from the nineteen migrant deaths that occurred recently in the tractor trailer in Victoria, Texas. More typically, in smuggling cases, law enforcement officers will encounter a small group of persons along the border with a "coyote," the person designated to guide migrants over the border. In these situations, law enforcement officials must act immediately to identify suspects, identify possible witnesses or victims, and gather evidence before it is destroyed, or the trail goes cold. Some investigations develop based on intelligence derived from sources such as informant tips, information from foreign authorities, or evidence obtained during the investigation of other crimes. This type of investigation may be long-term, resource-intensive, and involve the use of cooperating individuals and/or undercover agents.

In the trafficking context, law enforcement officers are often tipped off by escaped victims and non-governmental groups. Such tips may be communicated directly to an officer on the street, or may be made through a confidential phone call to the Trafficking in Persons and Worker Exploitation Task Force's Complaint Line at the Department of Justice. Sometimes raids of illicit facilities (such as brothels) may result in the discovery of trafficking victims.

Prosecutors and agents in smuggling and trafficking cases face difficult issues, not the least of which is deciding what to do with the affected migrants. Some may return, voluntarily or through removal, to their home country, while others may remain in the United States to assist in the investigation or prosecution. Trafficking victims may also apply to receive T non-immigrant status, if they are assisting to investigate or prosecute human traffickers. Prosecutors must develop leads and evidence to effectively prosecute smugglers and traffickers, while conserving scarce resources and avoiding inadvertently encouraging or "rewarding" illegal immigration.

<u>Coordination With Foreign Law Enforcement</u>

By definition, alien smuggling and international trafficking in persons involve more than one country.  Every international criminal law investigation poses distinct challenges, and, depending on the country involved, some may be more difficult than others.  United States law enforcement authorities operating abroad must always be cognizant of issues involving sovereignty and the application of foreign laws, and must also be prepared to deal with the capabilities and limitations of local law enforcement authorities in the foreign country.

Individuals associated with terrorist organizations have been known to use existing smuggling organizations and document vendors to facilitate their travel in various parts of the world.  Consequently, the Department of Justice, in concert with the Department of Homeland Security (including the Bureau of  Immigration and Customs Enforcement, the Bureau of Customs and Border Protection, and the Coast Guard), are moving aggressively to minimize threats to our national security.  One such effort is to target major smuggling networks that pose a particular threat to national security.

<u>Reorganization Of Resources Within The Justice Department</u>

In 2000, the Department of Justice created the Alien Smuggling Task Force within the Criminal Division.  Recently, the Department determined that it should institutionalize and expand the work of the Alien Smuggling Task Force, and in December 2002, the Task Force was incorporated within the newly-created Domestic Security Section of the Criminal Division. Cases involving sex trafficking of juveniles are handled by the Criminal Division's Child Exploitation and Obscenity Section.  The Civil Rights Division's Criminal Section, in conjunction with the United States Attorneys Offices, prosecutes trafficking cases involving sweat shops, domestic servitude, agricultural workers, and brothels. The Department, working

with United States Attorney's Offices around the country, has mounted an intensified effort in

recent years to combat this inhumane and often violent conduct. This intensified effort resulted

in the creation of the multi-agency Trafficking in Persons and Worker Exploitation Task Force,

begun in 1999.

<u>Interagency and International Cooperation</u>

To investigate and prosecute these offenses, the Criminal and Civil Rights Divisions

work collaboratively with other U.S. agencies, such as with the FBI, the Department of

Homeland Security, the Labor Department, and the State Department, as well as with foreign

authorities. We also have worked, as part of the interagency community, to strengthen the laws

of other countries and to make smuggling and trafficking extraditable offenses. We have assisted

other countries in their efforts to remove corrupt officials who aid alien smuggling organizations,

and have helped foreign counterparts to initiate their own prosecutions against targets deemed

significant by the United States. We believe that the United States Government must continue to

expand our international immigration enforcement efforts.

<u>Significant Prosecutions</u>

I would like to discuss just a few examples of alien smuggling and human trafficking

prosecutions the Department has brought in the last few years. Last year, here in the District of

Columbia, the Department prosecuted an Iranian national, Mohammed Hussein Assadi, for alien

smuggling. He was convicted following a jury trial in October 2002. Assadi ran a large

organization that smuggled aliens, generally from Middle Eastern or South Asian countries, into

the United States via commercial airlines. Assadi's ring provided aliens with stolen, photo-substituted passports from those European countries that qualify for visa waiver privileges.  As a result of interagency and international cooperation from other countries, Assadi was apprehended in the United States after being expelled from a foreign country.  He ultimately was sentenced to 30 months.

Another successful prosecution, which was brought by the U.S. Attorney's Office in the Southern District of Texas, involved Kenny Feng and other members of his organization.  Feng was a Taiwanese smuggling organizer, often referred to as a "snakehead," whose organization assisted in smuggling Chinese migrants to Latin America by boat.  Feng affiliated with other smugglers to transport migrants from China to the coast of Guatemala, where the human cargo would be offloaded and held in Guatemalan safe houses pending payment of smuggling fees.  Those who paid the fee would be transferred to other smugglers, who then would bring them overland from Guatemala to the United States.  The family of one female migrant paid $15,000 to be smuggled into the United States.  Upon her arrival in Guatemala,  however, the woman learned that her fee had been raised to $40,000.  The victim was then held in Guatemala for more than fifteen months.  Ultimately, the woman was sold to Mexican smugglers, who brought her into Texas.  While being held in Houston by members of the Mexican organization, the woman broke her back during an attempt to escape through a second-story window.  From her hospital bed, the woman cooperated with law enforcement.  As a result of law enforcement efforts, Feng was expelled from a foreign country, and then arrested in the United States.  Ultimately, a number of persons in the organization were apprehended and convicted for criminal offenses, including alien smuggling and hostage taking.  The length of sentences varied, with the longest being a 27-year sentence.

Just last week, on July 18, 2003, five Ecuadoran nationals pled guilty in the District of Columbia to conspiring to violate U.S. immigration laws. This prosecution arose out of the U.S. Coast Guard interdiction of two migrant smuggling vessels en route from Ecuador to Guatemala. The first was an unseaworthy fishing vessel with approximately 235 Ecuadorian migrants aboard. The second was a cargo vessel with 270 Ecuadorian and Indian migrants aboard. Food and water were limited, and the vessels were still days from reaching shore. As a result of the assistance of Ecuadorian and Mexican authorities, together with the close cooperation of the U.S. Coast Guard and the former Immigration and Naturalization Service (now the Bureau of Immigration and Customs Enforcement), this case was successfully prosecuted in the United States. The defendants face likely sentences of approximately three years.

The Department has also actively prosecuted human traffickers. Since January 2001, the Department of Justice has charged, convicted, or secured sentences for 106 human traffickers in 32 cases. The Department and USAOs have prosecuted seventeen cases involving 46 traffickers under statutes created in the TVPA.

Specific cases of note include *United States v. Kil Soo Lee*, which I mentioned earlier, involving sweatshop labor. In that case, Department of Labor, INS, and FBI investigators worked tirelessly and collaboratively to investigate a trafficking case involving 200 Vietnamese and Chinese nationals, mostly young women, whom the defendants brought from Vietnam to American Samoa to work as sewing machine operators in a Daewoosa garment factory in American Samoa. The victims, some of whom were held for up to two years, were forced to work through extreme food deprivation, beatings and physical restraint. The victims were held in barracks on a guarded company compound, and were threatened with confiscation of their passports, deportation, economic bankruptcy, severe economic hardship to family members, false arrest, and a host of other punishments and abuses. One victim had an eye gouged out by a defendant who struck her with a jagged pipe in order to punish her for refusing to comply with the defendants' orders. On February 21, 2003, a jury convicted Lee, the owner of the factory and leader of the operation, on nearly all counts. Two other defendants pled guilty. Lee faces a substantial prison term. Sentencing is expected in December 2003.

Several cases illustrate the inroads made by the Department in combating trafficking of domestic servants. In *United States v. Satia and Nanji*, two defendants were convicted of holding a teenage Cameroonian girl in involuntary servitude and illegally harboring her in their home to use her as their domestic servant. These criminals were each sentenced to 108 months in prison and ordered to pay $105,306.64 restitution to the victim. The defendants were convicted of involuntary servitude, conspiracy, and harboring the victim for their own financial benefit. The defendants recruited the girl to the United States with false promises of attending a U.S. school. Once she arrived here, she was isolated in the defendants' home and forced through threats, sexual assaults, and physical abuse to work for them for several years as their personal servant.

Recently, in *United States v. Blackwell*, the Department prosecuted three defendants, including a high-level Ghanaian government official, for tricking a young Ghanaian woman into working as a domestic servant in their home, under abominable working conditions. As a result of this important prosecution, two defendants were convicted. The Ghanaian government official remains in Ghana; however, she was stripped of the Cabinet post she held in her native country.

The Department's efforts in combating trafficking in farm workers is exemplified by *United States v. Lee*, in which three defendants pleaded guilty to rounding up and enslaving homeless and drug-addicted African-American men in Fort Pierce, Florida, and forcing them to pick oranges against their will by threats and violence, and by using crack-cocaine as a reward.

In the realm of sex trafficking, the Department made ground-breaking strides in *United States v. Jimenez-Calderon*, where the Civil Rights and Criminal Divisions, in collaboration with the United States Attorney's Office for the District of New Jersey, secured one of the first convictions under 18 U.S.C. § 1591. In this case, five defendants were charged on September 26, 2002, with conspiring to lure and transport young Mexican girls into the United States under false pretenses, and then forcing them into prostitution, using physical violence and threats to maintain strict control over them. In addition to the five indicted defendants, three other traffickers entered guilty pleas to sex trafficking charges in September and October 2002. Earlier this year, on the eve of trial, three of the five remaining defendants entered guilty pleas to various offenses, including conspiracy, sex trafficking by force, fraud or coercion, and conspiracy to obstruct justice. The final two defendants are fugitives.

<u>Sentencing And Penalty Issues</u>

The Department realizes the importance of sentencing issues in alien smuggling and trafficking in persons cases. The Department is taking steps to ensure the Sentencing Guidelines are applied as intended, while still taking into account the difficult issues faced by the large number of alien smuggling cases handled by some districts, particularly those along the Southwest Border. In trafficking cases, we have obtained significant penalties. In *United States v. Tecum*, the defendants, husband and wife, used fraudulent identity documents to smuggle a young Guatemalan woman into the United States through Arizona. Defendant Jose Tecum persuaded her to live as his wife, despite the fact that he already had a wife in Florida, and forced her to perform both housework and agricultural labor in California and Florida in order to pay off her smuggling debt. The defendant was convicted on charges of involuntary servitude, alien smuggling, kidnapping and document fraud and sentenced to 108 months in prison. In *United States v. Lakireddy*, a sex-trafficking case involving the largest and wealthiest landlord in Berkeley, California, who trafficked young girls and women into the United States from India for his sexual gratification, the lead defendant was sentenced to 97 months in prison and ordered to pay $2 million in restitution to his victims. Trafficking in persons are some of the most time- and labor-intensive cases the Department faces. Particular prosecutorial and sentencing challenges arise in the context of consent of aliens to be smuggled into the United States, but then they end up in an exploitative sexual or labor situation. Nonetheless, as detailed above, we have received significant sentences for traffickers.

We have also obtained significant sentences in a number of smuggling cases, particularly in those egregious crimes that result in injury or death. For example, earlier this year, Ruben Patrick Valdes, head of a smuggling organization who specialized in bringing large numbers of persons into the United States in tractor-trailers, and whose operation resulted in at least two known deaths, received a 27-year sentence. In another significant case, in 2002, Jorge Aleman, a ruthless smuggler whose dangerous maritime smuggling operation between Cuba and the United States resulted in at least one death, was convicted in the Southern District of Florida and sentenced to life imprisonment. When being chased by Cuban border guards, Aleman's practice was to throw migrants overboard so that the guards were forced to curtail the chase in order to rescue the migrants in the water.

In general, the sentences in most alien smuggling cases are much less, particularly where the United States does not have specific evidence to establish that the smuggler knew that the persons smuggled were criminals or where the migrants were not subject to significant risk of injury or death. For Fiscal Year 2001, of the 1,900 convicted alien smuggling defendants, approximately 87 percent of alien smuggling defendants received prison sentences, and the average sentence for those who went to prison was sixteen months. In human trafficking cases, in Fiscal Year 2001 there were thirteen defendants, ten of whom were convicted. Of these, four received a prison terms and six received probation. The average prison term for the four cases was 128.5 months with a range of 60-240 months. For the six that received probation, the average probation term imposed was 30 months, ranging from twelve to 36 months.

We have learned from experience that trafficking cases involve enormous commitments of time and resources. These cases often involve large numbers of victims, language barriers, multiple investigating agencies, international investigations and, frequently, victims who have suffered severe physical and psychological trauma. Despite such obstacles, the Department continues to bring these cases in greater numbers than ever before, having tripled the number of prosecutions in FY 2002 and FY 2001 as compared to the previous two fiscal years. Fortunately, the TVPA created new protections for trafficking victims that allow us to ensure that they are treated as victims and not as criminals. Victims can now qualify for refugee services and, in appropriate instances, immigration relief in the form of T non-immigrant status that allow victims to remain in the United States for three years. The coordination of victim services has demanded time, resources and commitment, particularly in cases involving large numbers of victims.

I note that the Trafficking Victims Protection Reauthorization Act (TVPRA), introduced in the House, includes jurisdictional fixes to 18 U.S.C. § 1591 which should assist our ability to prosecute sex trafficking cases. We will monitor sentencing results and work closely with Congress and the Sentencing Commission to ensure that appropriate sentences result for heinous traffickers.

Further, it is my understanding that the Sentencing Commission has put on its agenda for the upcoming year a review of immigration-related guidelines. The Criminal Division looks forward to working with this Subcommittee and the Sentencing Commission on these issues. In particular, as noted above and as will be highlighted in the testimony of United States Attorney Charlton, we believe that the sentences in alien smuggling cases do not always appropriately reflect the seriousness of the crime. We also look forward to working with the Subcommittee on review, improvement and potential expansion of our current alien smuggling laws. We believe that such a review could assist the government's efforts to discourage illegal migration generally, and to prosecute smugglers specifically.

I would be happy to answer any questions that the Subcommittee might have.

APEX
TP_CF
*9780160713286*